Persistence and Spacetime

Persistence and Spacetime

Yuri Balashov

OXFORD
UNIVERSITY PRESS

OXFORD
UNIVERSITY PRESS

Great Clarendon Street, Oxford OX2 6DP

Oxford University Press is a department of the University of Oxford.
It furthers the University's objective of excellence in research, scholarship,
and education by publishing worldwide in

Oxford New York

Auckland Cape Town Dar es Salaam Hong Kong Karachi
Kuala Lumpur Madrid Melbourne Mexico City Nairobi
New Delhi Shanghai Taipei Toronto

With offices in

Argentina Austria Brazil Chile Czech Republic France Greece
Guatemala Hungary Italy Japan Poland Portugal Singapore
South Korea Switzerland Thailand Turkey Ukraine Vietnam

Oxford is a registered trade mark of Oxford University Press
in the UK and in certain other countries

Published in the United States
by Oxford University Press Inc., New York

British Library Cataloguing in Publication Data

Data available

Library of Congress Cataloging in Publication Data

Library of Congress Control Number: 2009943746

Typeset by Laserwords Private Limited, Chennai, India
Printed in Great Britain
on acid-free paper by the
MPG Books Group, Bodmin and King's Lynn

ISBN 978-0-19-957992-1

10 9 8 7 6 5 4 3 2 1

For Marianna

Acknowledgements

I acknowledge support from the Willson Center for the Humanities and Arts and the University of Georgia Research Foundation, in the form of grants and teaching releases, which has allowed me to devote more time and effort to this project than would be possible otherwise. I extend my thanks to the Royal Institute of Philosophy, the London School of Economics, Institute for History and Foundations of Science at the University of Utrecht, and philosophy departments at Duke University, Idaho State University, and the University of South Carolina for opportunities to give colloquium and workshop presentations related to this project.

In my work on this book I have benefited greatly from numerous discussions with my colleagues. I would like to thank the following people for helpful comments: Bana Bashour, Darrin Belousek, Peter Bokulich, Carolyn Brighouse, Andrew Cortens, Chuck Cross, Jim Cushing, Mauro Dorato, Arthur Fine, Mark Heller, Don Howard, John Kennedy, Trenton Merricks, J. Brian Pitts, Oliver Pooley, Mike Rea, Robert Rynasiewicz, Ted Sider, James Simmons, Peter van Inwagen, Dean Zimmerman, participants in my seminars at the University of Georgia, and anonymous referees.

I am particularly grateful to Maureen Donnelly and Hud Hudson for extensive discussions and advice on the material presented in the book. Special thanks are due to Cody Gilmore who provided detailed and insightful comments on the entire manuscript.

I would also like to thank Peter Momtchiloff for his encouragement and my research assistant Charles Hollingsworth, the copy-editor Angela Anstey-Holroyd, and my editors at Oxford Catherine Berry and Louise Sprake, for their help at various stages of the work on the manuscript.

For permission to reuse portions of previously published material, I would like to acknowledge the following: 'Persistence and Space-Time: Philosophical Lessons of the Pole and Barn,' *The Monist* 83 (2000): 321–40, Copyright (c) 2000, THE MONIST: An International Quarterly Journal of Philosophical Inquiry, Peru, Illinois, USA 61354; 'On Stages, Worms, and

Relativity,' in Craig Callender (ed.), *Time, Reality, and Experience* (Cambridge: Cambridge University Press, 2002), 223–52; 'Special Relativity, Coexistence and Temporal Parts: A Reply to Gilmore,' *Philosophical Studies* 124 (2005): 1–40, with kind permission of Springer Science and Business Media; 'On Vagueness, 4D and Diachronic Universalism,' *The Australasian Journal of Philosophy* 83 (2005): 523–31, with permission from Taylor and Francis; 'Defining "Exdurance",' *Philosophical Studies* 133 (2007): 143–9, with kind permission of Springer Science and Business Media; 'About Stage Universalism,' *Philosophical Quarterly* 57 (2007): 21–39; 'Persistence and Multilocation in Spacetime,' in D. Dieks (ed.), *The Ontology of Spacetime*, vol. 2 (Amsterdam: Elsevier, 2008), 59–81, with permission from Elsevier; 'Pegs, Boards, and Relativistic Perdurance,' *Pacific Philosophical Quarterly* 90 (2009): 167–75. Some portions of the content of Chapters 1, 2, 4, and 5 will also be appearing in 'Persistence,' in Craig Callender (ed.), *Oxford Handbook of Time* (Oxford: Oxford University Press, forthcoming).

Contents

List of Figures

Introduction

This book investigates the implications of Einstein's theory of relativity for the debate about persistence. The contemporary approach to spacetime theories[1] has proven very useful in exploring the range of implications that physical theories, such as special and general relativity, may have for the philosophy of time. Their significance for the debate about persistence, on the other hand, has not been widely appreciated yet. It seems natural to think that a major transformation in our views of space and time associated with the theory of relativity should be highly relevant to that debate. But relativistic considerations very seldom figure in it. And when they do, their import is often unclear. As a result, the debate largely proceeds as if there were no relativity theory on the market. This is not to say that there are no exceptions. A few valuable discussions of persistence in the relativistic context have appeared in recent years.[2] But their weight in the overall dialectic of the debate is small. The present study seeks to remedy that situation.

My overall conclusion—that relativistic considerations favor four-dimensionalism over three-dimensionalism—is hardly surprising. It is, however, anything but trivial. Contrary to a fairly common misconception, there is no straightforward argument from relativity to four-dimensionalism. In this book I go even farther and defend the view that there is no deductively valid argument to this effect, straightforward or not. All the competing doctrines of persistence can be *stated* in the relativistic frame-work without violating their spirit. When so stated, however, they do not fare equally well. Making this clear requires more work than one might initially expect. And the final result hinges, not on demonstrative inference, but on a detailed cost-benefit analysis. But many interesting philosophical arguments are of this sort.

---

[1] Developed in seminal works by Anderson 1967; Friedman 1983; Torretti 1983; Earman 1989; and many others.

[2] Rea 1998; Balashov 1999, 2000a, 2000b, 2000c, 2002, 2003a, 2003b, 2005c, 2008, 2009; Sider 2001: §4.4; Gilmore 2002, 2006, 2007, 2008; Hales and Johnson 2003; Miller 2004; Gibson and Pooley 2006; Hudson 2006: ch. 5; Sattig 2006: §5.4; Eagle 2009a and 2009b.

Moreover, in deciding whether the theoretical costs of a particular position are acceptable, it is not enough, in this case, to appeal to common intuitions, for many such intuitions have been shattered by the advent of relativity. One should, therefore, give special attention to scientific considerations, and this to a greater extent than may be necessary in dealing with certain other metaphysical topics. The present study is intended to do just that: to put physics-inspired arguments ahead of many others. Consequently, the scope of this study is restricted and the modal force of its conclusions limited. But no harm is done. Scientific considerations are important, while the boundary between metaphysics and science is vague. The methodology of this project stems from the belief that current mutual separation, even alienation, of philosophical sub-disciplines (and of philosophy and science in general) is a regrettable state of affairs and that the resulting low level of communication between the corresponding communities and frequent mutual misunderstanding of their respective agendas is unprofitable. Facilitating dialogue is therefore a pressing matter.

The plan of the book is as follows. Chapter 1 introduces various assumptions adopted throughout the book and delineates its scope. Chapter 2 describes the major modes of persistence—endurance, perdurance, and exdurance—in a generic spacetime framework. Chapter 3 is an informal introduction to the geometrical structure of classical and special relativistic spacetime. This chapter can be skipped without harm by anyone familiar with the subject. The framework of Chapter 2 is then adapted, in Chapter 4, to classical spacetime. There I also take issue with a popular argument in support of four-dimensionalism, the argument from vagueness, and consider the implications of this critique for mereological universalism, a view about composition that is popular among four-dimensionalists.

Chapter 5 takes up the task of situating the rival views of persistence in Minkowski spacetime of special relativity. Along with defending my own strategy, I critically discuss alternative contemporary approaches and some earlier relativity-inspired arguments for four-dimensionalism. In Chapter 6 I introduce and defend a non-trivial notion of coexistence in Minkowski spacetime, which is then put to work, in Chapter 7, to support perdurance against endurance and exdurance. Chapter 8 develops and defends a different relativistic argument favoring perdurance over endurance, based on the relativity of shapes and other related configurations in Minkowski spacetime.

1

Background and Assumptions

This chapter introduces the framework for the subsequent discussion. Of special note are various assumptions adopted throughout the book. Some of them are more controversial than others and continue to be actively debated. There is a great deal to be said about each of them. Although I attempt to make my reasons for adopting such assumptions explicit and, where possible, note the implications of abandoning them, considerations of space prevent me from providing extended arguments in their defense. Some of these topics are revisited in later chapters.

1.1. Persistence and Philosophy of Time

The current mature stage in the debate about persistence was preceded by a period when the issue of persistence was not clearly disentangled from related but distinct issues in the philosophy of time.[1] The former is about the ontological nature of physical objects, while the latter are primarily about the nature of time.

For the purpose of this study, adopting a certain view of the nature of time is a prerequisite for a fruitful discussion of the ontology of persistence. The view of time presupposed here is inspired, to a large extent, by modern science and can be described as *spacetime realism*. It combines several elements, which are best illustrated by considering some contrasts.

1.1.1. *Eternalism v. presentism*

On a popular characterization of eternalism, this doctrine holds that all moments of time and their contents enjoy the same ontological status. Past

---

[1] A vestige of such entanglement is found in the ambiguity of the term 'four-dimensionalism' used to denote both a certain ontology of persistence and a certain ontology of time, or spacetime.

and future moments, events, and objects are just as real as the present ones; they are just not "temporally here." (Compare: the planet Uranus is not spatially here, but it does not, for that reason, fail to exist.) Eternalism contrasts with presentism, the view that only the present entities are real. Eternalism and presentism are widely regarded[2] as the best contemporary incarnations of the older, and more broadly defined, competing views of time known as A- and B-theories, or tensed and tenseless theories.[3]

Along with the ontological claim that all moments of time and their contents are equally real, eternalism incorporates the linguistic claim that tensed locutions, such as 'Gail will fail,' have tenseless truth conditions expressed by metalinguistic token-reflexive clauses such as 'Gail (tenselessly) fails later than this utterance of "Gail will fail" ' or by sentences incorporating dates into their content: 'Gail (tenselessly) fails on July 25, 2009.'[4] Presentism, on the other hand, must locate the truth conditions of tensed sentences in tensed facts about the present state of the world. It is not immediately clear what these facts could be, and the issue figures prominently in the contemporary dispute about the pros and cons of presentism and eternalism.[5]

I mention these issues only to set them aside. Eternalism is the only position available to anyone sharing my approach to contemporary spacetime theories, whereas presentism is strictly inconsistent with it. The approach adopted in this book regards physical theories, such as classical mechanics and special relativity (SR), as based on distinctive claims about the intrinsic geometry of the spacetime manifold. Since the geometry of Minkowski spacetime does not support a frame-invariant notion of simultaneity, it does not allow one to define the concept of *the present*. And without such a concept, presentism cannot get off the ground.[6] To put the point

---

[2] By many contemporary philosophers of time, but not by all. Dissenters include Broad 1923; Smith 1993; Tooley 1997; and McCall 2004.

[3] The labels 'tensed/tenseless theories' and 'A/B-theories' are still used today to pick out certain combinations of linguistic and ontological doctrines. Some authors (but not all) consider the labels as interchangeable. There does not seem to be general agreement on the usage of this terminology in the literature.

[4] For a recent discussion of these two approaches, see Smart 2008.

[5] For recent contributions see Bigelow 1996; Zimmerman 1998a and 2008; Markosian 2003; Crisp 2005 and 2007; Bourne 2006; and references therein.

[6] This seems obvious to many, but it may require support, for one might argue that the doctrine of presentism could be made compatible with the lack of absolute simultaneity by modifying its letter without abandoning its spirit (see, e.g., some contributions to Craig and Smith 2008 and Zimmerman forthcoming). Various ways in which this could be done are discussed and shown to be untenable in Savitt 2000; Callender 2000a; Sider 2001: §2.4; Saunders 2002; and Balashov and Janssen 2003.

vividly, the presentist is committed to the following: when I click my finger on Betelgeuse Boris Yeltsin is either alive or dead. But according to special relativity, there is simply no fact of the matter. There is no global present moment cutting throughout the entire universe that has more than a frame-relative significance.

Or is there?

1.1.2. *Interpretations of SR*

Some authors have contended that facts about absolute simultaneity and the absolute present have a place in SR after all, provided this theory is given a suitable "neo-Lorentzian" reinterpretation, and argued that this reinterpretation is physically acceptable and metaphysically preferable to the standard formulation.[7]

The idea is that one could accept all the empirical consequences of this theory (including length contraction, time dilation, and so on) and yet insist that there is a privileged inertial reference frame, in which meter sticks *really* have the length they do and time intervals between events refer to the *real* time. Associated with this reference frame would be a set of hyperplanes uniquely foliating spacetime into equivalence classes of absolutely simultaneous points. Such a privileged reference frame would not be distinguished in any empirical sense and would not be identifiable in any real experience. Thus the speed of light measured in any inertial frame would still be exactly c, the number obtained by dividing the *apparent* distance covered by light by the *apparent* time spent. The point of introducing the notion of a reference frame privileged in this sense would be to draw an ontologically important distinction while preventing any observer from discovering the difference.

This strategy—which is, often with little warrant, associated with the name of H. A. Lorentz—is misguided both philosophically and historically. There is a sense in which an approach of this sort is on a par with proposals to return to the days before Darwin in biology or the days before Copernicus in astronomy. Arguing this point in detail, however, could well take another book-length study.[8] Here I simply note that I put "neo-Lorentzian"

---

[7] See, in particular, Craig 2001; some contributions to Craig and Smith 2008; and Zimmerman forthcoming.

[8] Perhaps worth pursuing, given the popularity of the approach in some circles. Some of my views of the matter are expressed in Balashov and Janssen 2003. For a more extended methodological argument

interpretations of SR aside. But advocates of such interpretations could still accept my arguments as conditional on the adoption of the standard spacetime approach.

1.1.3. *Manifold realism v. relationism*

But what is the "standard spacetime approach?" Its roots are often traced to Hermann Minkowski's seminal work (Minkowski 1908) which introduced the notions of spacetime manifold (*Mannigfaltigkeit*), of wordlines in spacetime representing careers of idealized physical objects, and of the intrinsic geometry of the manifold neatly expressed in its "lightcone structure." It should be noted that Einstein's original 1905 formulation of SR had a very different mathematical form and did not incorporate any of these notions. After some initial hesitation Einstein accepted Minkowski's geometrical approach. It proved valuable in Einstein's own path to general relativity and greatly facilitated the understanding of SR. On the philosophical front, the geometrical approach was later developed into a highly successful framework for a unified treatment of various local spacetime theories.[9] With the advent of scientific realism, it became only natural to view this success story in light of ontological commitment.[10] Spacetime geometry came to be regarded as being explanatorily effective.[11]

None of this, of course, guarantees the truth of the story. One could deny the ontological significance of the unity of space and time (of which more below) reflected in the notion of the spacetime manifold and view their unification merely as a mathematical tool. Or one could ascribe reality to the unity but deny the existence of spacetime as an entity separate from physical bodies and events. This stance has a venerable history. To illustrate, suppose that two objects are 5 meters apart at a given moment of time (speaking classically). The *substantivalist* will say that the objects bear this distance relation derivatively, by occupying independently existing spacetime points (or regions), which stand in this relation fundamentally. The *relationist*, on the other hand, kills the middleman and

against theories such as "neo-Lorentzian relativity," based on the analysis of *common-origin inference*, see Janssen 2002*a*. For an authoritative historical account of the "Lorentz-Einstein controversy," see Janssen 2002*b*.

   [9] See Friedman 1983; Torretti 1983; Earman 1989.
   [10] See, in particular, Smart 1963, 1972; Nerlich 1994.
   [11] For a sustained attack on this view, see Brown 2005. For a recent critique of Harvey Brown's program, see Norton 2008*a*.

insists that the objects bear the distance relation directly. *Mutatis mutandis* for events.

For many, philosophers and non-philosophers alike, substantivalism is the default view and relationism is a revisionary view. Relationism confronts the task of reformulating the whole of physics without postulating an independently existing spacetime background.[12] On the other hand, relationism has recently derived support from the "hole argument."[13] The argument, in its own turn, has not gone unchallenged.[14] The present state of the controversy between substantivalism and relationism remains complex and this brief note cannot begin to do justice to it.[15]

This study adopts spacetime substantivalism. Here I join other participants in the debate about persistence who, with very rare exceptions (and for the most part, without making it explicit), presuppose this view. It is unclear how to state the positions in the debate without invoking the relation of *location*, or *occupation* of spacetime regions by objects, and it is unclear whether the notions of location and occupation could be adequately captured in relationist terms.[16] Far from suggesting that a substantive claim about the ontological nature of spacetime should be accepted simply because all or most accept or presuppose it, it is important to emphasize that every inquiry has its limits and that the implications of rejecting substantivalism for the entire agenda of contemporary metaphysics may be far more dramatic than many participants realize.

1.1.4. *Space, time, and spacetime*

The notion of spacetime is central to this project. It is also indispensable to much of contemporary physics. This notion, however, is at some level of abstraction from the ordinary way of thinking about the physical world. In our everyday experience we normally represent an event as something that happens at a certain place and at a given time, where ascribing a spatial location to the event is taken to be independent of its characterization

---

[12] A program pursued by several physicists and philosophers. See, in particular, Barbour 1999; Belot 2000; Brown and Pooley 2002; Huggett 2006.

[13] For a useful introduction to the hole argument industry, see Norton 2008*b*.

[14] See, e.g., Maudlin 1990.

[15] We also abstract from *supersubstantivalism*, the view that spacetime is the only entity in existence. Some implications of supersubstantivalism for the persistence debate are discussed in Sider 2001: §4.8. For related discussions, see also Skow 2007; Parsons 2007; and Schaffer 2009.

[16] See Hawthorne and Sider 2002 for a recent discussion of how far such a program could be pursued.

as something that happens at a particular time. Intuitively, one conceives of points and regions of space as somehow "enduring" through time. According to the spacetime view, this conception is fundamentally wrong and, perhaps, not even meaningful. Spacetime cannot, in general, be constructed by taking space and simply "adding" time to it. Rather, one should start with spacetime and then—if really needed, and under certain conditions—"extract" space and time from it.

Spacetime is a more intimate union of space and time than a simple combination (or a "Cartesian product") of them. There is a sense in which the members of this union do not exist independently of each other. There are various ways to make this sense precise, and some of them will be explored later on. To get a flavor of it,[17] think of a moment of time as a global flat spacelike *slice* through spacetime, and think of a position in space as a timelike *line* extending throughout the spacetime manifold. Then moments of time and positions in space literally *overlap* and therefore do not exist independently of each other.

1.2. Atomism and Composition

1.2.1. *Atomism*

This study presupposes that spacetime and matter are atomic or "pointilliste."[18] Although this assumption may seem very natural, it is controversial. First, the idea that all material objects are composed of simple objects that do not have proper parts has been challenged by the advocates of "atomless gunk."[19] Second, the idea that mereological atoms, if they exist, must be pointlike has been called in question by the defenders of extended simples.[20] Third, the idea that atomism has a firm grounding in classical science, from which contemporary metaphysics has always derived its main inspiration (the picture of "atoms in a void"), has been challenged by recent in-depth studies of the history and foundations of analytical mechanics.[21] Fourth, the

[17] The flavor comes from Gilmore 2004: 23.
[18] The latter term coined by Jeremy Butterfield 2006*b*.
[19] For an influential defense of gunk, see Zimmerman 1996*b*. For recent discussions, see a special volume of *The Monist* devoted to simples and gunk (Hudson 2004).
[20] See, in particular, Parsons 2000; Simons 2004; Hudson 2006: ch. 4; and McDaniel 2007.
[21] Butterfield 2006*b*.

idea that the fundamental ontological category is that of *things* has recently been questioned and a radically new approach based on the category of *stuff* has been developed.[22] On top of all this, the idea that space, time, and spacetime may not themselves be atomic—and thus should not be thought of as being composed of dimensionless points and moments—has recently been advanced by a number of authors.[23]

Adopting atomism against this backdrop is undoubtedly controversial, but it allows me to avoid issues that are tangential to this project. I hope it will, in the end, become clear that most of the issues central to it are not sensitive to taking a stance on "atomic pointillisme," while being able to use its resources greatly simplifies discussion. In that respect, the assumption of atomism can be looked upon as an idealization.[24]

1.2.2. *Composition*

In light of the previous assumption, every material object can be looked upon as being composed, ultimately, of pointlike objects. This, by itself, does not tell us whether, and under what conditions, two or more objects, pointlike or not, compose a further object, or have a fusion. The topic of composition occupies a central place in present-day metaphysics and substantially overlaps the debate about persistence. Two popular views on composition shared by many (albeit different) participants in the persistence debate mark the extreme sides of the spectrum: nihilism and universalism. According to (mereological) nihilism, *no* plurality of objects has a fusion; in other words, no composite objects exist. In conjunction with atomism this means that the only existing objects are pointlike atoms. The opposite view known as (mereological) universalism holds that *every* set of objects has a fusion. In other words, any plurality of objects composes a further object. There is, for example, an object exactly composed of my left ear and the polar star. The two extreme doctrines allow plenty of room for "intermediate" views, on which some, but not all sets of objects have a fusion.[25]

---

[22] Markosian 2004*a* and MS; Kleinschmidt 2007.

[23] See, in particular, Forrest 1996; Arntzenius 2008; and Russell 2008.

[24] This said, some popular arguments figuring in the persistence debate (but not considered in this book) may be sensitive to adopting or rejecting atomism. One important example is the "rotating disk argument" against perdurantism. As argued by Butterfield 2006*a*, abandoning "pointillisme" takes the sting out of this argument.

[25] A prominent example of an intermediate view is Peter van Inwagen's theory (1990*b*), which recognizes two kinds of objects; mereological atoms and a special category of their composites: living beings.

Much of what follows tacitly assumes an "intermediate" view. I hasten to note that, for the most part, this assumption will not be supported by any arguments. Such a view, however, seems to me reasonable and it comes to bear on what I have to say about persistence in special relativistic spacetime in Chapters 5–8, although to a varying extent. Much of Chapters 5–7 operate with composite material objects but could, at a great loss to vividness, be restated in terms of atoms. Chapter 8, on the other hand, relies on interestingly shaped composite material objects (boxes, spheres, etc.) and its substance would be affected if it turned out that no such objects existed or if they existed on a par with widely scattered objects.

My stance on composition becomes more explicit in Chapter 4, where I consider the relationship between four-dimensionalism and mereological universalism. It turns out that many of those who subscribe to the former also endorse the latter.[26] I regard this as an interesting fact about a certain community, but unlike many others, I do not think that adopting four-dimensionalism provides a strong motivation for endorsing universalism or that endorsing the latter on independent grounds provides a good argument for the former. In §§4.2–4.6 I take issue with those who defend these connections. My approach there is based on drawing an important distinction between two types of composition: achronal and diachronic, and the corresponding distinction between two types of universalism.

1.3. Scope

The scope of the ensuing discussion and arguments is, for the most part, restricted to classical (non-quantum) SR. This has the advantage of making the discussion manageable. But it certainly brings with it serious limitations. Although Minkowski spacetime is a good approximation of most of the spacetime of our world, it is, in the end, just that: an approximation.

It is not immediately clear how to extrapolate the notions central to the debate about persistence to general relativistic spacetime, which has no place for global moments of time and intuitive analogs of "momentary locations." More importantly, it is even less clear how to think about

---

[26] Examples include Quine 1960: 171; Heller 1990; Hudson 2001: ch. 3 and 2006: ch. 5; and Sider 2001.

persistence in the context of (non-relativistic) quantum mechanics and it is entirely unclear how to begin thinking about it in the context of quantum field theory. Indeed, it is unclear how to apply endurantist concepts (such as 'being wholly present at multiple spacetime regions') even to classical fields. This is not to suggest that questions of this sort should not be raised. But their serious discussion requires extensive preliminary work. I am hoping that pursuing my limited agenda will go some way towards fulfilling the prerequisites for more ambitious projects.

1.4. Some Matters of Methodology

The methodological strategy of this project is to bring physical considerations to bear on an important metaphysical problem. This requires taking a stance on the controversial issue of the relationship between physics and metaphysics.[27] But it also has more immediate consequences.

One of them concerns an ongoing controversy about defining the terms central to the debate about persistence, such as 'being wholly present' and 'temporal part.' Much of the controversy surrounding attempts at such definitions is inspired by counterexamples that are physically impossible, or at least remote, even if they are possible in a broader sense (i.e. logically or metaphysically). Since my central concern is to find out how various views of persistence square with the physics of the actual world the scope of my discussion is obviously limited to the realm of the physically possible and I can implement this program without considering physically impossible scenarios. This enables me to use simpler definitions and not worry about the problems they might create in a broader context. No doubt, this closes some doors (and perhaps eliminates part of the excitement from the enterprise). But it allows me to focus on the chosen agenda and avoid orthogonal engagements.

Giving scientific considerations prominent weight in a metaphysical discussion such as this may also impact the issue of how much weight should be given to other, more traditional considerations based on common beliefs and conceptual analysis. This makes the discussion both easier and more difficult. Easier, because giving priority to considerations of a certain sort

---

[27] For recent critical discussions of this issue, see Monton MS and Eagle MS.

tends to simplify the overall analysis. More difficult, because it often involves swimming in uncharted waters. There is no escape from ordinary beliefs: every inquiry, no matter how abstract and technical, must be anchored in them at some point. But abandoning the friendly confines of the classical worldview for the highly counterintuitive relativistic setting makes the very identification of safe "anchoring points" part of the problem. This problem will surface many times in the ensuing discussion.

Finally, this work presupposes that the metaphysical disagreement between different views of persistence is real and not merely linguistic or descriptive. The world of three-dimensional enduring entities, which are capable of multilocation, is fundamentally different from the world of four-dimensional temporally extended spacetime "worms." Or so it appears. Relatedly, the world in which only present entities and moments of time exist is fundamentally different from the world in which all moments of time and their contents are equally real.

But the idea that there is, in both cases, genuine metaphysical dis-agreement has recently been attacked in a number of works defending a broadly Carnapean—skeptical and deflationary—approach to problems in fundamental ontology. This, in turn, prompted a series of increasingly sophisticated replies. As a result, what was initially a side issue grew rapidly into a flourishing industry of *metaontology*.[28]

While fully aware of the new skeptical challenge, I do not address it explicitly in this book. The topic of metaontology is complex and requires special attention. I also cannot hope to convert the skeptic by simply pursuing my agenda. I do hope that when the issues central to this project are laid down in some detail, it will become clear that the main action takes place sufficiently far from the metaontological frontlines to allow the serious ontologist at least temporary refuge. And what more can one expect?

---

[28] Chalmers, Manley, and Wasserman 2009 is a state-of-the-art collection, which brings together contributions from both sides of this divide.

2

Persistence, Location, and Multilocation in Spacetime

2.1. Endurance, Perdurance, Exdurance: Some Pictures

Atoms and molecules, desks and computers, dogs, butterflies, and human persons persist through time and survive change. This much is obvious. But *how* do material objects manage to do it? What are the underlying facts of persistence? This is currently a matter of intense debate. And it is a relatively recent debate. Some forty years ago most philosophers would not recognize the question of persistence as deserving more than casual attention. And when they did, the issue would quickly get boiled down to some combination of older themes. Here is my dog Pif, and there he is again. He changed in-between from being calm to being angry. But what is the big deal? Things change all the time without becoming distinct from themselves (as long as they do not lose any of their essential properties, some would add). Is there anything more to persistence? Have we made any progress since Aristotle?

We have. Today we know that there is much more to say. The problem of persistence has become, in the first place, a problem in *mereology*, a theory of parts and wholes.[1] It has also become an issue in a theory of *location*.[2] These two topics continue to drive the debate.

What parts do persisting things have? Most people are happy to concede that a table, say, has four legs and a top; each of these, in turn, has smaller

---

[1] For an authoritative exposition of classical mereology, see Simons 1987.

[2] For a systematic unified treatment of the topics of parthood and location, see Casati and Varzi 1999.

parts. A dog is similarly composed of a myriad of cells. These *spatial* parts constitute the selfsame entity at any moment of its existence.[3] That does not preclude mereological change. The dog, in particular, retains his identity through continuous replacement of his cells. This is, by and large, the ordinary picture of persistence.

But there are dissenters anxious to revise it. They note that most physical objects are spatially composite but insist that composition takes place along the temporal dimension too. Just as the dog has distinct spatial parts, it also has different temporal parts at different moments of its existence. Moreover, the dog's spatial parts also have temporal parts, and vice versa. One can slice up an object first in a spatial direction and then in the temporal direction, or one can do it in a reverse order, and end up with the same—much as you can cut a cake first lengthwise and then across or vice versa. Thus today's temporal parts of the head, the tail, the legs, and the torso compose today's temporal part of Pif *le chien*. None of these objects, big or small, existed yesterday, and none will exist tomorrow. Yesterday's temporal part of Pif was made of numerically distinct parts of his spatial constituents. The dog persists by being composed of non-identical temporal dog-parts, much as a road persists through space. And just as the road changes from being narrow in the country to being wide in town, the dog changes from being calm to being angry by having parts—temporal parts—that are calm and those that are angry. Change is qualitative variation in space for the road, and in time for the dog. Ditto for other physical objects.

The views sketched above are variously known as *three-dimensionalism* (3Dism), or *endurantism*, and *four-dimensionalism* (4Dism), or *perdurantism*. They disagree about the manner in which physical objects persist over time and through change. Some of the disagreement boils down to the question of what parts persisting objects have. Perdurantists typically insist, while endurantists typically deny, that objects have temporal as well as spatial parts.

In addition, these theorists typically disagree about where and how objects are located in spacetime. The locus of a perduring object is a

---

[3] Such parts are spatial in the obvious sense that they are distributed in the three dimensions of space. It is true that salient spatial parts of ordinary objects are usually identified, not by their occupying particular spatial regions, but by other features, e.g. by their functional role, or relations in which they stand to other such parts and to the whole. This does not detract from the spatial nature of such parts, although it may give the designation 'spatial' a somewhat unnatural flavor. Be that as it may, the designation is widely used in the literature and we shall adopt it below.

four-dimensional (4D) region of spacetime which, intuitively, incorporates the object's entire career. A 4D perduring object exactly fits in its spacetime career path and is only partially located at what we normally take to be a region of space at a certain moment of time, a three-dimensional (3D) slice of spacetime. A 3D enduring object, on the contrary, occupies its spacetime path in virtue of being wholly present (i.e. present in its entirety, with no part being absent) at multiple instantaneous slices of spacetime. The key idea here is multilocation: one object—many locations.

This leaves room for another view known as *stage theory*, or *exdurantism*, which seeks to combine certain features of endurantism and perdurantism. Like perdurantism, stage theory endorses the existence of temporal parts, or stages. But like endurantism (and contrary to perdurantism), stage theory typically identifies ordinary objects with 3D entities that are wholly located at momentary regions that lack temporal extension. Despite their temporal shortness, such entities—object stages—persist. They manage to do so by *exduring*—by standing in temporal *counterpart* relations to later and earlier object stages. (The reader will note a close analogy with modal realism.)

These pictures are rough and ready. The real situation is more complex. To begin with, there is no general consensus on how to *state* the rival views of persistence and what exactly is at stake in the debate.[4] This has resulted in crisscrossing taxonomies and continual redrawing of the boundaries that were previously regarded as fixed. Furthermore, the persistence debate is closely entangled with a number of other philosophical disputes, old and new, which are equally complex: the nature and ontology of time, parts and wholes, material constitution, personal identity, modality, causation and properties, reference, and vagueness (the list is hardly complete). Equally important, considerations from physics have an important bearing on the issue of persistence.

Insofar as the rival views on persistence can be stated with sufficient clarity to the satisfaction of the warring parties, endurantism tends to enjoy the default advantage of a "common sense view," while perdurantism and exdurantism have a characteristically "revisionist" flavor and, as a consequence, are expected to do more work to motivate and defend themselves. This may or may not be fair, but the situation is not uncommon: in many other metaphysical disputes, there are "default" positions (e.g.

---

[4] For a particularly incisive recent analysis of these issues, see Hawthorne 2006.

presentism in the philosophy of time, dualism in the philosophy of mind, Platonism in the philosophy of mathematics) and their theoretical rivals (respectively: eternalism, materialism, and nominalism) whose appreciation requires serious ontological commitment, skepticism about the scope of traditional conceptual analysis, and sometimes a deferential attitude towards scientific evidence.

Be that as it may, there has been no shortage of arguments in favor of 4Dism and against 3Dism.[5] Many of them are driven by philosophical reflection on the problems of change, intrinsic properties, temporal predication, material constitution, and vagueness. There has been no shortage of 3Dist responses to those arguments.[6] The ensuing discussions continue to benefit not only direct participants, but neighboring areas mentioned above. Several excellent review articles cover these grounds in detail.[7]

But persistence is a dynamic and rapidly evolving topic, and the debate is changing every day. As already mentioned, recent developments have tended to focus on the set of related issues having to do with parthood and location.[8] It is in this context that broadly empirical considerations are increasingly brought to bear on the discussion.[9] The bulk of the present book is devoted to this decidedly positive tendency. We shall approach this task gradually, starting with a review of more traditional themes and another informal tour of the whole territory.

2.2. More Pictures

Consider a material object, such as a proverbial poker, that persists over time and changes from being hot at midnight to being cold the next day. As

[5] See, in particular, Armstrong 1980; Balashov 1999, 2000a, 2000b, 2000c, 2002, 2005b, 2009; Effingham and Robson 2007; Hawley 2001; Heller 1990; Hudson 2001; Lewis 1983, 1986: 202–4, 1988; Sider 2001, 2008.

[6] See, in particular, Baker 2000; Haslanger 1989, 2003; Johnston 1987; Lowe 1988; McGrath 2007a; Merricks 1994, 1999; Oderberg 1993; Rea 1998; Sattig 2006; Thomson 1983; van Inwagen 1990a; Zimmerman 1998a.

[7] See Haslanger 2003; Hawley 2008; McGrath 2007b; Sider 2008. Haslanger and Kurtz 2006 is a useful anthology of recent work on persistence over time.

[8] See, in particular, Bittner and Donnelly 2004; Crisp and Smith 2005; Hudson 2006; Gilmore 2006, 2007, 2008, 2009a, 2009b; Sattig 2006; Parsons 2007; Balashov 2008; Donnelly 2009; Eagle 2009a, 2009b; Saucedo forthcoming; Kleinschmidt forthcoming.

[9] See Rea 1998; Balashov 1999, 2000a, 2000b, 2000c, 2003a, 2003b, 2005c, 2008, 2009; Sider 2001: §4.4; Gilmore 2002, 2006, 2007, 2008, 2009b; Hales and Johnson 2003; Miller 2004; Gibson and Pooley 2006; Eagle 2009a, 2009b.

already noted, the parties to the debate agree that the poker has a spacetime *career* represented by a 4D *path*—the shaded region in Figures 2.1(a–c).[10] But they disagree about the manner of the poker's location at its path. The endurantist will say that the poker is multilocated at 3D *slices* of its path corresponding to different times; call them '*t*-slices.' To pull off this trick, the poker must fit, in its entirety, into every such slice and must, therefore,

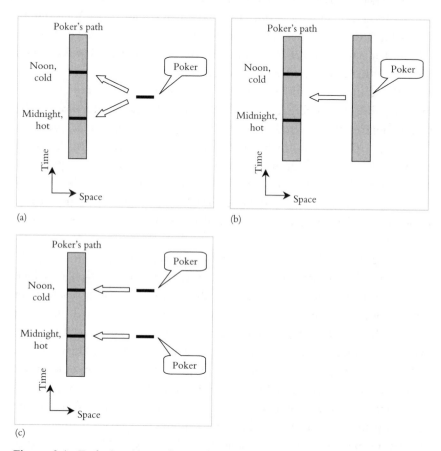

Figure 2.1. Enduring (a), perduring (b), and exduring (c) poker in spacetime.

---

[10] It is difficult to draw full-blown 4D diagrams. Accordingly, we shall follow others in suppressing one or two dimensions of space, as in Figure 2.1. The only exception will occur in §8.3. (Those who are curious may want to jump, for a moment, to Figure 8.9 (on p. 203) for a sneak preview of a real 4D diagram.) Of course, one should never suppress the time dimension. Figures 2.1(a) and 2.1(b) are inspired by Gilmore 2006: 205.

be a 3D object. The perdurantist, on the other hand, will say that the poker is singly located at its path and is, therefore, a 4D object.

In the present context, location means *exact location*. The idea is that a spacetime region at which an object is exactly located is a region into which the object exactly fits and which has exactly the same size, shape, and dimensionality as the object itself. Exact location has a number of notable *negative* properties. If object *o* is located at each of a number of distinct spacetime regions, it does *not* follow that *o* is located at their union. If *o* is located at a proper subregion R of a larger region R′, it does *not* follow that *o* is located at R′. And conversely, if *o* is located at R, it does *not* follow that *o* is located at any proper subregion of R.[11]

Some of these features of the relation of exact location are crucial to understanding the difference between endurance and perdurance. If the poker endures it is (exactly) located at various *t*-slices of its path but not at any other region, including the path in its entirety. On the other hand, if the poker perdures it is exactly located at its entire path but not at any other region, including various *t*-slices of its path. What is located at such slices are distinct temporal parts ('*t*-parts') of the poker. Endurance and perdurance are easy to describe and visualize. What about exdurance?

If the poker exdures it is exactly located at some *t*-slice of its path. More carefully, what is located at this slice is the poker *t*-stage, which is a poker in its entirety: no parts of the poker are missing from it. In other words, an exduring object is wholly present at a single moment of time. In this respect, exdurance is analogous to endurance. One may even be inclined to depict exdurance in a way similar to Figure 2.1(a). But that would be wrong. Figure 2.1(a) represents the relation of location, which holds between an enduring object and many instantaneous regions of spacetime at which it is exactly located (i.e. multiple *t*-slices of the object's path). But an exduring object cannot be a relatum in this sort of relation, because it is not capable of multilocation. The poker *t*-stage is located *only* at the corresponding *t*-slice of the poker's path. What is

---

[11] For a more detailed discussion of the properties of exact location, see Gilmore 2006: 200–4. The above concept of exact location or its analogs are increasingly used in the literature on persistence. See, in particular, Bittner and Donnelly 2004; Hudson 2001 and 2006; Balashov 2008; and Donnelly 2009. It should be noted, however, that other ways of understanding the basic concept of location are possible. See, in this connection, Parsons 2007; Gilmore 2008 and 2009a; Eagle 2009a; Saucedo forthcoming.

located at another slice is a numerically distinct poker stage. All such stages bear to each other a temporal counterpart relation characteristic of pokers. This relation is extensionally equivalent to one that would hold between the corresponding temporal parts of the poker, if the latter perdured. But since it does not—since, by assumption, the poker exdures rather than perdures—what occupies each t-slice of the poker's path is a poker, not a poker part. The distinctive features of exdurance are somewhat elusive, and it is not so clear how to represent them in a diagram. Figure 2.1(c) may suggest helpful directions, especially when it is viewed against Figures 2.1(a) and 2.1(b).

But there is also a useful non-pictorial analogy. In David Lewis's ontology of concrete *possibilia* each object has modal counterparts in many distinct worlds, and each counterpart represents a way an actual object might have been. Our talk of various counterfactual situations involving, say, Hubert Humphrey is underwritten by Humphrey's distinct modal counterparts located in separate worlds. Similarly, when we talk about what the exduring poker does at different times, this talk is underwritten by distinct poker stages wholly confined to those times.

This helps to highlight the difference between endurance and exdurance. Exdurantists generally accept, but endurantists generally deny, the existence of momentary object stages—entities *exclusively* confined to t-slices of the objects' paths in spacetime. They will, however, agree that such a 3D entity *represents* a persisting object in its entirety—directly or vicariously, as the case may be. This sets them apart from perdurantists, who insist that the relevant persisting object is only partly present at a t-slice of its path, while being much longer temporally. Note that exdurantists need not deny the existence of such longer entities, which are aggregates of distinct t-stages. That is why the title '4Dism' is appropriate for exdurance, as well as perdurance. What the exdurantist must resist is the identification of the long entity with an ordinary persisting object. She will say that a dog is a short dog-stage, not a long spacetime dog-worm. The difference between perdurance and exdurance thus appears to be semantical not ontological, as both parties typically accept instantaneous material stages, as well as their cross-temporal aggregates, in their ontologies. As we shall see shortly, the "merely semantical" difference is difference enough.

But as a preliminary step, we need to correct an unfortunate feature of the above description, which makes exdurance needlessly handicapped right from the start. The offending feature has to do with the failure of exduring objects to be multilocated in spacetime. But on a common understanding of persistence, something persists only if it exists at more than one moment,[12] and an instantaneous object-stage, strictly speaking, does not. One could, of course, simply accept this consequence and conclude that exduring objects do not persist. Indeed, such considerations have been adduced to disqualify exdurance from being a legitimate mode of persistence. But in my view, pursuing this line would be unfair, especially in light of certain theoretical benefits that might be uniquely associated with exdurance.[13] At the definitional stage, "vicarious persistence" via counterparts should be as respectable as "direct persistence."[14]

What we need is a generalization of the notion of exact location, on which both endurance and exdurance would exemplify *some* mode of multilocation. The generalized notion of exact location will be introduced more formally in §2.4 below. Before we do it, we need to discuss another consequence of the informal pictures of endurance, perdurance, and exdurance, which has to do with the attribution of temporary properties.

2.3. Temporal Modification and the "Problem of Temporary Intrinsics"

2.3.1. *Temporal modification*

To say what properties (and spatial parts) a persisting object has at a moment of time, the endurantist who subscribes to spacetime realism must relativize possession of temporary properties (and spatial parts) to time.[15] She cannot say that the poker is hot and stop here, because the selfsame poker is also cold, when it is wholly present at a different time. Time must somehow be worked into the picture. One has to explain how time interacts with the attribution of properties and, relatedly, what makes statements ascribing

---

[12] The locus classicus is probably Lewis 1986: 202, "Something *persists* iff, somehow or other, it exists at various times."

[13] Such benefits are thoroughly discussed in Sider 2001: 188–208 and Hawley 2001: chs. 2 and 6. Balashov 2005*b* defends exdurance as the best way to account for certain features of the experience of time. But see Moyer 2008 for a critique of "slice theory" (his name for the exdurance theory).

[14] The situation is rather similar to that surrounding Kripke's "Humphrey objection" to modal realism.

[15] Indeed, the need to do so may be taken as definitional of endurance. See Hawley 2001: 27.

temporary properties to objects true. There are several ways of doing it, which bring with them somewhat distinct metaphysics and semantics of temporal modification.[16]

On *relationalism*,[17] the enduring poker comes to be hot at midnight and cold at noon by bearing (tenselessly) the relations *hot-at* and *cold-at* to midnight and noon, respectively. Besides relationalism, distinct variants of the endurantist relativization strategy include *indexicalism* and *adverbialism*.[18] Instead of making a moment of time a full-fledged relatum of a relation which generates a temporary property, the first variant makes time modify the *property* itself and the second the *having* of it. According to indexicalism, the correct analysis of the poker's being hot at midnight and cold at noon proceeds by welding the times with the properties and producing two seamless time-indexed items *hot-at-midnight* and *cold-at-noon*, which are then attributed, tenselessly or atemporally, to the poker. On one version of adverbialism, the poker manages to be hot at midnight and cold at noon by having the simple property hotness in a certain *way*, the midnight way, or *midnightly*, and the simple property coldness in a different way, the noon way, or *noonly*.

According to perdurance theory, on the other hand, the poker is a 4D entity extended both in space and time. It persists by having distinct temporal parts at every moment of its existence. When we say that the poker is hot at midnight and cold at noon, what we really mean is that the poker's midnight part has the former property and its noon part the latter. The sense in which the properties of the poker's temporal parts can be attributed to the 4D whole is, in many ways, similar to the sense in which the properties of the spatial parts of an extended object are sometimes attributed to the whole. When we say that the oil pipe is hot in the vicinity of the pump and cold elsewhere, we really mean that the pipe has, among its spatial parts, a part in the vicinity of the pump, which is hot, and a part elsewhere, which is cold. Just as the pipe (and entire thing) *changes* from being hot to being cold along its spatial extension, the poker (the entire

---

[16] The general strategy of relativizing temporary properties to times was sketched by Lewis 1986: 202–4. It was then developed by others, in a number of works and in different forms. For recent contributions and references see MacBride 2001 and Haslanger 2003. But see Sattig 2006 for objections to such relativization strategies.

[17] Not to be confused with spacetime *relationism*. See §1.2.3 above.

[18] Neither term is standard or universally accepted, but both are widely used (and sometimes confused with relationalism) in the literature.

perduring object) changes from being hot to being cold along its temporal dimension.

Thus the perdurantist should analyze possession of temporary properties by temporally extended objects in terms of exemplification of such properties *simpliciter* by the objects' temporal parts, or stages. What about the exdurantist? The latter will agree that the midnight and noon poker stages are hot and cold *simpliciter*. In difference from perdurance theory, however, the exdurantist will deny that such stages are temporal *parts* of the poker (or in any event, *proper* temporal parts of the poker); rather each of them *is* the poker at a corresponding moment of its existence. Figure 2.1(c) suggests one way of thinking of it by allowing the label 'Poker' to have multiple referents, one per each moment of time (consider, by analogy, the referential behavior of an indexical, such as 'now'); *mutatis mutandis* for other singular terms. At different times, 'Poker' refers to different poker stages, which are the *same* poker, where sameness should be construed, not as numerical identity, but as a temporal counterpart relation, C_p.[19]

As already noted, the exdurance theory of persistence and temporal modification strikes many as being very problematic. So it is appropriate to devote a bit more time to establishing its initial credentials. Pick one of the poker's tonight stages, poker$_1$. Poker$_1$ is a familiar ordinary object, the poker, and is hot. This object will persist until tomorrow and will be cold then. One worry comes from skepticism that a *momentary* object could, in principle, accomplish this feat. Stage theory's answer is that poker$_1$ persists by exduring and comes to possess the *temporal* property *will be cold* by being C_p-related to poker$_2$, a certain poker stage tomorrow, which is cold *simpliciter*. There is no doubt that the stage view's reliance on the counterpart account of persistence and temporal modification presents a theoretical cost. But it also provides for important theoretical benefits—a point already made (see note 13). Theories must be judged by carefully balancing their benefits over the costs, not on the basis of their proximity to common sense. Again the analogy with Kripke's "Humphrey objection" to modal realism should be kept in mind.

[19] On "sameness through time" as distinct from sameness of identity, see Hawley 2001: 62 ff. Sameness through time in stage theory is distantly similar to the sameness of two books read by different people. The analogy is distant because the latter sameness is sameness of abstract types, while the former is supposed to be sameness of concrete objects.

How short are object stages? Although the view that they have non-zero duration could perhaps be coherently developed,[20] it is more natural to require that stages be instantaneous, to accommodate, in the simplest possible way, continuous change of spatial position and of other instantaneous properties.[21] There is a sense then in which ordinary objects—tennis balls, cats, and persons—are, on stage theory, very short-lived. As noted above, this does not prevent them from persisting (courtesy the counterpart view) and possessing many familiar momentary properties, such as temperature, color, shape, or position. We have just seen that stages can also possess historical and future-oriented properties, such as *was hot* and *will be cold*. But at first sight, the instantaneous nature of object stages appears to prevent them from having "lingering" properties (Hawley 2001: 54), which take time to be instantiated. Examples include *orbiting the Earth*, *speeding*, *traveling across the tennis court*, *getting wet*, *dreaming of the Bahamas*, and *writing War and Peace*. A single object stage can *be* wet, but it does not seem capable of *getting* wet, for that requires being first dry and then wet, and no instantaneous entity can be both. Similarly, orbiting the Earth *now* involves having certain properties in the past and future.

But since attribution of past and future properties to present stages does not constitute a problem for stage theory, neither should the ascription of lingering properties. My current stage can be getting wet by being covered with water and being counterpart-related to earlier stages that are dry and to later stages that are covered with more water. Single object stages can have lingering properties by standing in appropriate relations to surrounding stages. Lingering properties, on this view, are highly relational, but they are, nonetheless, the properties of instantaneous objects. And this, upon reflection, seems to be the right result. Moreover, it is a familiar result. Many physical properties, such as velocity and acceleration, are instantaneous, but their possession by objects at single instants is partly a matter of what goes on at other instants. There are also useful spatial analogies: an array of shingles comes to possess the property of being a roof by being appropriately related to other parts of the house.[22] An isolated array of shingles does not have this property. Despite being, in this sense,

---

[20] See Butterfield 2006a for a recent proposal. The view that all stages are temporally extended could then be combined with the view that time itself is "gunky" (if the latter view could be coherently developed). On gunky space and time, see Forrest 1996; Arntzenius 2008; and Russell 2008.

[21] See Hawley 2001: 50 and Sider 2001: 197–8.   [22] Cf. Hawley 2001: 65 and Sider 2001: 197–8.

relational, the property in question is possessed by a single array of shingles, for a single such array need not be isolated. Similarly, an isolated stage cannot be getting wet. But a single stage can, for single stages need not be isolated.

In general, object o at t (i.e., a momentary object stage) has a lingering property Φ_L in virtue of (i) having intrinsic features pertinent to instantiating Φ_L at t and (ii) bearing C_o to object stages at times earlier and later than t, where such stages have certain intrinsic features pertinent to o's instantiating Φ_L at t and C_o is a counterpart relation unifying the object stages in question.

2.3.2. The "problem of temporary intrinsics"

The need to relativize properties or their possession to times, in order to account for change, was often portrayed, in earlier discussions of persistence, as a distinctive disadvantage of endurance.[23] The idea is that many of the variable properties of persisting object—length, shape, temperature, texture, and so forth—are "temporary intrinsics": they characterize ways objects are at a certain moment of time "in and of themselves," not in relation to anything else. This demand is allegedly satisfied by the perdurance theory, where possession of temporary properties by temporally extended objects is analyzed in terms of exemplification of such properties *simpliciter* by the objects' temporal parts, but not by the endurance theory, which, as we have seen, must resort to one of the relativization schemes.[24]

But what exactly is wrong with relativization of properties, or their instantiation, to times? Philosophers who are inclined to reject this move tend to do so on the grounds that paradigm temporary properties, such as shapes, are not relations. Lewis, in particular, urged that relativization eliminates the having of familiar properties *simpliciter* in favor of having them always in temporally modified ways (Lewis 1986: 204 and 1988: 65)

---

[23] See, in particular, Lewis 1986: 202–4.

[24] The *presentist* endurantist escapes the need for relativization because the only properties the object has are those it has at the present moment. See, in this connection, Merricks 1994; Hinchliff 1996; and Zimmerman 1998a. Some authors have argued that this mode of exemplification of temporary properties *simpliciter* favors the combination of presentism and endurance over any other ontology of time and persistence. Those who espouse this approach sometimes go further and argue that endurance *entails* presentism. See, in particular, Merricks 1995 and 1999. In light of the problems befalling presentism this entailment may be looked upon as needlessly burdensome. Indeed, many endurantists are eternalists. See, in particular, Johnston 1987; Haslanger 1989 and 2003; van Inwagen 1990a; and Rea 1998. In any event, for the purpose of this project, presentism is not a live option. See §1.1.1.

and is, therefore, objectionable. But many others have responded by arguing that relations to times are different in nature from relations to other things and, therefore, relativization to time does not deprive temporary properties of their intrinsicality.[25] This response has merit. Furthermore, *pace* Lewis, it is unclear that perdurantism escapes the need for relativization either. Perduring objects do not have temporary properties *simpliciter*, but only in virtue of a parthood *relation* to their temporal parts that do.

It may begin to look as though endurance emerges as a winner here: in the ontology of stages, momentary properties are exemplified by persisting objects directly, for such objects are instantaneous stages, not aggregates thereof.[26] This advantage, however, is offset by the need (see §2.3.1 above) to bring in counterpart relations to neighboring objects stages to account for exemplification by instantaneous stages of historical and "lingering" properties, such as *was bent*, *growing*, and *getting tired*. As we have seen, these properties are highly relational, and their exemplification by object stages is not just a matter of having a certain quality *simpliciter* and *nothing else*.

Thus it appears that everyone must relativize one way or another; hence, the need to do so does not obviously put any theory of persistence at a disadvantage.

This concludes an informal sketch of three major views of persistence: endurantism, perdurantism, and exdurantism, and of their characteristic accounts of the attribution of temporary properties. We shall return to the issue of temporal modification in §4.1 and will consider its relativistic extension in §5.1.

Below, I attempt to make the distinctions among the three modes of persistence more precise and generalize them in a way that will allow us to situate the rival views of persistence in a generic spacetime framework.[27] This requires, among other things, replacement of classical notions such

---

[25] For recent discussions see MacBride 2001: 84; Hawley 2001: §1.4; and Haslanger 2003.

[26] The supremacy of stage theory vis-á-vis the problem of temporary intrinsics is discussed in Sider 2000.

[27] The spacetime approach adopted below has much in common with those developed in Rea 1998; Balashov 2000b; Sider 2001: 79–87; Hudson 2001 and 2006; Gilmore 2004 and 2006; Crisp and Smith 2005; and Sattig 2006. Some of my terminology and basic notions come from Gilmore 2004. Bittner and Donnelly 2004 offer a rigorous axiomatic approach to explicating the mereological and locational notions central to the debate about persistence. Their approach is set in a broadly classical context but can be usefully extended to the generic spacetime framework. This is accomplished in Donnelly 2009, which is, by far, the most comprehensive study of the formal properties of various region-relative parthood relations in spacetime.

as 'temporal part,' 'spatial part,' 'moment of time,' and the like with their more appropriate spacetime counterparts. The resulting generic framework will then be adapted to the more specific environment of Galilean and Minkowski spacetime in Chapters 4 and 5.

2.4. Persistence, Location, and Multilocation in Generic Spacetime

We shall begin by introducing some underlying spacetime concepts.

2.4.1. *Absolute chronological precedence*

We shall take the relation of *absolute chronological precedence* ($<$) as undefined. Informally, spacetime point p_1 stands in this relation to p_2 ($p_1 < p_2$) just in case p_1 is earlier than p_2 in every (inertial) reference frame.[28] It is natural to assume that absolute chronological precedence is asymmetrical ($p_1 < p_2 \rightarrow \neg p_2 < p_1$) and, hence, irreflexive ($\neg p < p$).

2.4.2. *Achronal regions*

Next we define the notion of an *achronal spacetime region*. A spacetime region (i.e., a set of spacetime points) is achronal iff no point in it absolute-chronologically precedes any other point.

(D2.1) Spacetime region R is *achronal* $=_{df} \forall p_1, p_2 \ (p_1, p_2 \in R \rightarrow \neg p_1 < p_2)$.

Achronal regions are three-dimensional "slices" through spacetime that generalize the classical notion of a moment of time. In fact, a moment of time could be defined as a *maximal* achronal region of spacetime with a certain property:

(D2.2) R is a *moment of time* $=_{df}$ (i) R is a maximal achronal region of spacetime; (ii) R is $\Omega =_{df} [(\forall p_1, p_2) \ [p_1, p_2 \in R \rightarrow \neg p_1 < p_2] \land (\forall p) \ ((\forall p_1, p_2) \ [p_1, p_2 \in R \cup \{p\} \rightarrow \neg p_1 < p_2] \rightarrow p \in R) \land R$ is $\Omega]$.

---

[28] As one would expect (§§4.1 and 5.1), in classical spacetime absolute chronological precedence can be taken literally to mean precedence in the absolute time while in the special relativistic framework absolute chronological precedence is equivalent to the frame-invariant relation in which two points stand just in case they are either (i) timelike separated or (ii) lightlike (null) separated while being distinct.

Clause (ii) is needed because nothing in the above definition requires an achronal region to be a *flat* 3D hypersurface in spacetime. But it is natural to suppose that no achronal hypersurface can represent a moment of time in the classical or special relativistic setting unless it is flat. In these settings, 'Ω' could be taken to be synonymous with 'flat,' where flatness is defined in the usual metric way.[29]

The significance of flat achronal hypersurfaces in special relativistic spacetime and their relation to the notion of time are issues that require more discussion and I shall return to them in §5.2. But they do not play any part in the general definitions of the different modes of persistence provided later in this section. What does play a central role in them is the notion of achronality and the underlying relation of absolute chronological precedence. My approach takes the second notion as a starting point to allow maximum generality. But in familiar contexts, it bears close relationship to other widely used concepts. Thus in many applications, a maximal achronal region is none other than a Cauchy surface—a spacelike hypersurface that intersects every unbounded timelike curve at exactly one point. But there is no need to invoke additional notions, such as 'spacelike' and 'timelike,' in a generic context where all the useful work could be done by 'achronal.'

We need, however, to make a brief digression to note a familiar problem with the concepts of 'absolute chronological precedence' and 'achronal,' which is brought to light by considering peculiar spacetimes possessing closed or "almost closed" timelike curves. For the purpose of this informal consideration, 'timelike' could be taken to be synonymous with 'non-achronal.' Closed timelike curves exist, for example, in Gödelian cosmological models of general relativity, but a flat "cylindrical" spacetime could serve as a useful toy model.[30] It is easy to see that there is a sense in which two "nearby" points p_1 and p_2 can stand in the relation of absolute chronological precedence ($p_1 < p_2$)—the sense obtained by tracing a non-achronal curve from p_1 to p_2 around the "cylinder." But there is also a sense in which they are not ($\neg p_1 < p_2$)—the sense obtained by tracing an achronal curve from p_1 to p_2 along a generatrix of the "cylinder."

---

[29] What about general relativity? Although it goes beyond the scope of this study it is worth noting that, except in very special cases (e.g. certain idealized cosmological models), the notion of a moment of time lacks any non-local meaning in general relativistic spacetime.

[30] Cf. Gilmore 2007, where a similar toy model is used to investigate the implications of time travel scenarios for the issue of persistence.

Accordingly, a certain region containing both p_1 and p_2 might be classified by (D2.1) as being both achronal and non-achronal. Situations of this sort figure prominently in the literature on time travel.[31]

Another problem arises in spacetimes having a "trouser" topology.[32] Points p_1 and p_2 belonging to different legs of the "trousers" do not bear any well-defined metrical relations to each other and, hence, are not related by <. But if p_1 precedes the merger by just a few seconds but p_2 is thousands of years away from it, there is some inclination to say that p_2 chronologically precedes p_1 (in the sense associated with '<').

Both problems could perhaps be alleviated by making the definition of 'achronal region' in the relevant sense *local*[33] and thus consistent with closed or "almost closed" timelike curves, and with the "trouser" topology. We shall abstract from such situations in what follows. This limitation is quite tangential to our main task—to capture the central features of the various modes of persistence in a spacetime setting by using an economical set of primitive notions. We shall assume, accordingly, that global maximal achronal regions are always available.

2.4.3. *Location and quasi-location*

Persisting objects are located at regions of spacetime. For our purposes, 'located at' means *exactly located*. Recall the guiding idea: the region at which an object is exactly located is the region into which the object exactly fits and which has exactly the same size, shape, and position as the object itself.[34]

I take 'located at R' to mean the same as 'wholly present at R,' but I set aside the question of whether the latter notion can be rigorously defined for objects having (achronal) parts.[35] The prospects of such a definition have been intensely debated.[36] My concerns at this point are, however, rather

---

[31] For recent discussions, see Gilmore 2006 and 2007, and Gibson and Pooley 2006: §5.

[32] See Gilmore 2006: 204, ns 19 and 20, who refers in this connection to Sklar 1974: 306–7.

[33] See Gilmore 2006: 209, n. 19, for one attempt to do it.

[34] This notion of exact location is similar to Hudson's notion of *exact occupation* (Hudson 2001), Bittner and Donnelly's notion of *exact location* (Bittner and Donnelly 2004), Gilmore's notion of *occupation* (Gilmore 2006), and other equivalents found in the recent literature. But see Parsons 2007 for a very different approach to understanding 'exactly located.'

[35] I briefly revisit the issue at the end of this section. For a definition of 'achronal part,' see (D2.6) below.

[36] See Rea 1998; Sider 2001: 63–8; McKinnon 2002; Crisp and Smith 2005; Parsons 2007; and references therein.

orthogonal to it, for I am interested in the underlying sense of 'located at R' applicable to (achronally) composite and non-composite objects alike, which any such definition must take as a starting point.

What is essential to my task is that there be a common notion of location—call it *quasi-location* (or *q-location*)—which is broad enough to incorporate the modes in which both enduring and exduring objects are capable of multilocation. To repeat, the sense in which an exduring object accomplishes this feat is similar to the sense in which a world-bound individual of the Lewisian pluriverse, such as Humphrey, can nonetheless be said to exist at (or, as Lewis put it, "according to") multiple worlds. To make the notion of q-location precise, let us start with (non-modal) counterparthood and stipulate that every object (enduring, perduring, or exduring) is a (non-modal) counterpart of itself. This is a natural assumption that does not impose any undue commitments on endurantism or perdurantism. The advocates of both theories could agree that every persisting object has an *improper* non-modal counterpart: itself—multiply located in the case of endurance, and singly located in the case of perdurance.[37] An object can then be said to be *q-located* (*quasi-located*) at a region R just in case one of the object's counterparts is strictly located there:

(D2.3) *o* is (exactly) *q-located* at R $=_{df}$ one of *o*'s (non-modal) counterparts is (exactly) located at R.

Note immediately that if *o* is located at R then it is also q-located there, but not vice versa.

The following definitions[38] help to align q-location more precisely with the notion of persistence.

(D2.4) Spacetime region *o* is the *path* of object *o* $=_{df}$ *o* is the union of the spacetime region or regions at which *o* is q-located.

(D2.5) *o persists* $=_{df}$ *o*'s path is non-achronal.

[37] For those who may be inclined to resist this usage of 'counterpart' as too stretched, a somewhat less elegant equivalent of (D2.3) is readily available:

(D2.3′) *o* is (exactly) *q-located* at R $=_{df}$ *o* is (exactly) located at R or one of *o*'s (non-modal) counterparts is (exactly) located at R.

[38] Adapted from Gilmore 2004: chs. 2 and 2006: 204 ff. But Gilmore might object to combining his definition of 'persists' with the broad sense of quasi-location. On his official view, as far as I can tell, exduring objects do not persist. See, however, Gilmore 2006: 230, n. 21, where he suggests a rather innocuous modification to his approach that would accommodate exdurance.

The advantage of (D2.1) and (D2.3)–(D2.5) lies in their ability to offer a *unified* account of persistence and multilocation, on which (i) enduring, perduring, and exduring objects persist in the same sense, and (ii) enduring and exduring objects are multilocated in the same sense. All parties can agree that endurance, perdurance, and exdurance are bona fide modes of persistence and, in particular, that exdurance is not a second-class citizen: exduring objects persist in the same robust sense as enduring objects do. This allows one to focus on the important question of how they manage to do so.

2.4.4. *Achronal and diachronic parts*

Next we need generalizations of the concepts of spatial and temporal part. We shall take a three-place relation 'p is a part of o at achronal region R' as a primitive.[39] The intuitive ancestor of this relation is the familiar time-relativized sense in which certain cells are part of me at one time but not at another. Where p, o, and R stand in this relation, we shall say that p is an *achronal part* of o at achronal region R and denote it with the subscript '$\perp$':

(D2.6) $p_\perp$ is an *achronal part* of o at achronal spacetime region R $=_{df} p_\perp$ is a part of o at R.

Diachronic parthood could then be defined as follows:[40]

(D2.7) $p_\parallel$ is a *diachronic part* of o at achronal spacetime region R $=_{df}$ (i) $p_\parallel$ is located at R but only at R, (ii) $p_\parallel$ is a part of o at R, and (iii) $p_\parallel$ overlaps at R everything that is a part of o at R.

The subscripts '$\perp$' and '$\parallel$' thus indicate that the relevant dimension of parthood—the achronal or the diachronic—is, respectively, "orthogonal" or "parallel" to the dimension of time.[41]

---

[39] This important decision has various ramifications that are discussed in more detail at the end of the present section. For now, let me note that R-relativized parthood is similar to the relation used by Hudson 2001, in developing his "partist" view of persistence but more restrictive than the latter (and thus closer to the familiar concept of temporary parthood), in that Hudson's notion relativizes parthood to arbitrary regions of spacetime whereas mine is limited to achronal regions. The formal properties of various R-relativized parthood relations have received systematic treatment in the work of Maureen Donnelly 2009.

[40] (D2.7) follows the general pattern of Sider's definition of 'temporal part' (2001: 59).

[41] These subscripts introduce certain syntactical complexity in the ensuing definitions. But I think the notation is useful, and the complexity is unavoidable in any event. One could perhaps try to pack

Note that neither *p* nor *o* need be "as large as" the achronal region R, in order to stand in the relation '*p* is a part of *o* at R.' All that could reasonably be required of the achronal extents of *o* and *p* at R is that the intersection of *p*'s path with R be "within" the intersection of *o*'s path with R:

(WITHIN) *p* is a part of *o* at achronal region $R \rightarrow p \cap R \subseteq o \cap R$.

This, of course, entails that both $p \cap R$ and $o \cap R$ are "within" R. Thus my hand is a part of me at a certain momentary location of my hand, at a momentary location of my body, and at a momentary location of the Solar system. Furthermore, if I am an exduring object my hand is a part of me at an achronal region at which neither I nor my hand are even "sub-located"—say, a region at which I *was* located at some moment ten years ago. In this case the job of grounding R-relativized parthood is done by the non-modal counterparts of the relevant objects. Finally, assuming perdurance, one of my cells at *t* (i.e. a global moment of time) is a part of me at my momentary location at *t* (i.e. at the location of my momentary *t*-part), but also a part of me at the momentary location of the Solar system at *t*.[42]

In contrast, the notion of diachronic parthood is more restrictive: if $p_\parallel$ is a diachronic part of *o* at achronal region R then $p_\parallel$ must "fit into" R exactly, although *o* may "overfill" R in virtue of having parts (both achronal and diachronic) at superregions of R.

In the subsequent discussion the generic relations of achronal and diachronic parthood, explicated in (D2.6) and (D2.7), are restricted to distinguished achronal regions—those "containing" (in a relevant sense) the objects involved in the relation. Such regions are *achronal slices* of the objects' paths.

(D2.8) $R_\perp$ is an *achronal slice* of R $=_{df}$ $R_\perp$ is a non-empty intersection of a maximal achronal 3D region with R $=_{df}$ $(\exists R^*) [(\forall p_1, p_2) (p_1, p_2 \in R^* \rightarrow \neg p_1 < p_2) \land (\forall p) ((\forall p_1, p_2) [p_1, p_2 \in R^* \cup \{p\} \rightarrow \neg p_1 < p_2] \rightarrow p \in R^*) \land R_\perp = R \cap R^* \land (\exists p) p \in R_\perp]$

---

it into the definienses instead, but at the double cost of making the latter difficult to process and also losing the graphic vividness associated with '⊥' and '∥'.

[42] One counterintuitive consequence of R-relativized parthood thus understood must be noted: *p* may be a part of *o* at an achronal region "not large enough" for *o*, provided that it is "large enough" for *p*. For example (WITHIN), as stated above, does not preclude me from being a part of my hand at a momentary location of my hand. A fully axiomatic treatment of R-relativized parthood would probably need to rule out such cases, perhaps by modifying (WITHIN). This would lead to complications that are best avoided in the present context.

More comments are in order.

(i) As defined by (D2.6) and (D2.7), achronal and diachronic parthood are not mutually exclusive. Indeed, diachronic parthood is just a special case of achronal parthood. In the case of both perdurance and exdurance, the diachronic part of any object at a t-slice of its path is equally its achronal part at that slice. Thus my diachronic part at the current t-slice of my path is also my (improper) achronal part at that slice.

(ii) However, there is a sense in which *proper* achronal and diachronic parthood are exclusive. If *proper parthood at achronal region* R is defined as *asymmetrical* achronal parthood at R:

> (D2.9) $p_\perp$ is a *proper achronal part* of o at achronal region R $=_{df}$ (i) $p_\perp$ is an achronal part of o at R, (ii) o is not an achronal part of $p_\perp$ at R,

then, if $p_\perp$ is a proper achronal part of o at some achronal slice $o_\perp$ of its path then $p_\perp$ is not a diachronic part of o at $o_\perp$, proper or not. The reason, roughly, is that $p_\perp$ is "smaller" than o at $o_\perp$ and thus cannot be a diachronic part of o at $o_\perp$.[43] For example, my hand is, on (D2.6), (D2.7), and (D2.9), a proper achronal part of me, but not a diachronic part of me, at the current t-slice of my path.

And if *proper diachronic parthood* is defined as *asymmetrical* diachronic parthood:

> (D2.10) $p_\parallel$ is a *proper diachronic part* of o at achronal region R $=_{df}$ (i) $p_\parallel$ is a diachronic part of o at R, (ii) o is not a diachronic part of $p_\parallel$ at R,

then, if $p_\parallel$ is a proper diachronic part of o at some achronal slice $o_\perp$ of its path then $p_\parallel$ is not a proper achronal part of o at $o_\perp$. The reason, roughly, is that being a diachronic part of o at $o_\perp$, proper or not, makes $p_\parallel$ "as large as" o at $o_\perp$ and, hence, not a proper achronal part of it at $o_\perp$. However, $p_\parallel$ and o will in general be *improper* achronal parts of each other at $o_\perp$.

---

[43] Here and below one should presuppose an R-relativized analog of Strong Supplementation:

(SS^R) If x is not a part of y at R then there is a part of x at R that does not overlap y at R,

to rule out spurious asymmetry between x and y, not grounded in any difference in their parts at R, that would lead to a vacuous satisfaction of the definiens of (D2.9). I thank Cody Gilmore for this observation.

On the other hand, if proper achronal and diachronic parthood at achronal region R are understood as follows:

(D2.9′) $p_\perp$ is a *proper achronal part* of o at achronal region R $=_{df}$ (i) $p_\perp$ is an achronal part of o at R, (ii) $p_\perp \neq o$;

(D2.10′) $p_\parallel$ is a *proper diachronic part* of o at achronal region R $=_{df}$ (i) $p_\parallel$ is a diachronic part of o at R, (ii) $p_\parallel \neq o$,

then one object could be both a proper achronal and a proper diachronic part of another object at some achronal region. Consider a perduring or exduring statue and the piece of clay of which it is composed. Some would argue that the statue (and hence, its *t*-part) is not identical with the piece of clay (and its corresponding *t*-part). If so then, by (D2.9′) and (D2.10′), the statue and the piece of clay are both proper achronal and proper diachronic parts of each other at the *t*-slice of the path of both objects.

(D2.9), (D2.10), (D2.9′), and (D2.10′) raise an interesting question of how to develop general R-relativized mereology, but we must leave the matter here.[44]

(iii) As defined, achronal and diachronic parts are achronal, that is, diachronically (or temporally, where this designation is appropriate) non-extended. In this I deviate from the authors who explicitly allow temporally extended temporal parts and make them do some useful work.[45]

2.4.5. *o-eligible achronal slices of o*

Finally, we need the notion of an *object-eligible achronal slice* of that object's path:

(D2.11) $o_\perp$ is an *o-eligible achronal slice of o's path* $o =_{df}$ either o itself or o's diachronic part at $o_\perp$ is q-located at $o_\perp$.

This notion is intended to be neutral among the different modes of persistence and between the different modes of location (i.e., location versus q-location). Why do we need this notion? Its full significance will not

[44] For a systematic analysis of R-relativized mereology, see Donnelly 2009, which is highly recommended.

[45] Butterfield 2006a, for example, employs non-achronal temporal parts to rebut, on behalf of the perdurantist, Kripke's "rotating disk" argument. That argument is thoroughly debated in Armstrong 1980; Zimmerman 1998b, 1999; Lewis 1999; Callender 2000b; Sider 2001: §6.5; Hawley 2001: ch. 3. For earlier discussions of finitely extended temporal parts, see Heller 1990 and Zimmerman 1996a.

emerge until Chapter 5, but some preliminary consideration may help motivate it. Initially one might think that the notion is redundant, for how could
any achronal slice of *o*'s path *fail* to be *o*-eligible, given that *o*'s path is a union
of the spacetime region or regions at which *o* is q-located (D2.4 above)?
Doesn't the definition of *o*'s path entail "achronal universalism," the thesis
(i) that any enduring object is located at *every* achronal slice of its path, (ii)
any perduring object has a diachronic part at *every* achronal slice of its path,
and (iii) any exduring object is q-located at *every* achronal slice of its path?

It is important to see that the entailment does not hold. Suppose *o*
endures or exdures and is q-located at each of a continuous family of
achronal regions forming its path, but at *no* other region (see Figure 2.2).
Then each member of this family is, quite trivially, an *o*-eligible slice of *o*'s
path. But no "crisscrossing" achronal slice of *o*'s path, such as $o_\perp$*, is *o*-
eligible. The same result holds, *mutatis mutandis*, for perdurance. Suppose *o*
perdures, and each of the continuous family of achronal slices of its path
hosts *o*'s diachronic part at that slice. All the same, this does not guarantee
that the same privilege must be accorded to the "crisscrossing" slice $o_\perp$*.
For all we know, $o_\perp$* may fail to contain a diachronic part of *o*.

Possibilities of this sort may initially escape attention because they turn
on the existence of "crisscrossing" achronal slices, which are available in
some spacetimes (e.g. in Minkowski spacetime of special relativity), but
not in others. In particular, their absence from the classical spacetime may
constitute a good reason to embrace a version of "achronal universalism"

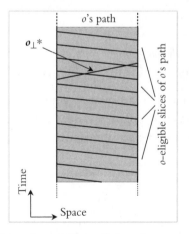

Figure 2.2. Crisscrossing achronal slices of *o*'s path.

(see §4.1), thus completely trivializing the notion of an o-eligible achronal slice. On the other hand, the pervasive presence of "crisscrossing" achronal slices in Minkowski spacetime raises some controversial questions about o-eligibility, which will be taken up in Chapter 5.

This is, of course, jumping too much ahead. The point to keep in mind at this stage is that it would be desirable not to read any debatable views into the generic notions of endurance, perdurance, and exdurance.

2.4.6. 'Endurance,' 'perdurance,' and 'exdurance' defined

The following definitions capture the important distinctions among the three modes of persistence.

(D2.12) o endures $=_{df}$ (i) o persists, (ii) o is located at every o-eligible achronal slice of its path, (iii) o is q-located only at o-eligible achronal slices of its path.

(D2.13) o perdures $=_{df}$ (i) o persists, (ii) o is q-located only at its path, (iii) the object located at any o-eligible achronal slice $o_\perp$ of o's path is a proper diachronic part of o at $o_\perp$.

(D2.14) o exdures $=_{df}$ (i) o persists, (ii) o is located at exactly one region, which is an o-eligible achronal slice of its path, (iii) o is q-located at every o-eligible achronal slice of its path.

On these definitions, the difference between endurance and perdurance is as expected: (i) enduring but not perduring objects are multilocated (and, hence, multi-q-located) in spacetime; (ii) perduring but not enduring objects have diachronic parts.[46]

More importantly, the definitions also bring out the crucial distinction between perdurance and exdurance: (a) exduring but not perduring objects are multi-q-located in spacetime; (b) while both perduring and exduring objects have diachronic parts, perduring objects have only *proper* diachronic parts.[47] That exduring objects have improper diachronic parts follows from

---

[46] Barring certain exotic exceptions; see below.

[47] It might seem odd to give quasi-locational properties so much weight in the definitions of the modes of persistence. In particular, why not replace q-location with location in clause (iii) of (D2.12) and clause (ii) of (D2.13)? (Thanks to Cody Gilmore and a referee for raising this question.) The reason to keep them in place is to make q-location do useful work in *strengthening* the definitions of endurance and perdurance. Clause (iii) of (D2.12) requires that, not only an enduring object, but any of its non-modal counterparts be, in effect, temporally unextended. And clause (ii) of (D2.13) precludes, not only a perduring object, but any of its non-modal counterparts, from being located anywhere except at the object's path.

clauses (ii) and (iii) of (D2.13) and the definition of 'diachronic part,' which together entail that the object located at every achronal slice of an exduring object's path is a diachronic part, at that slice, of some object: namely, itself.[48]

Finally, the definitions pinpoint the difference between exdurance and endurance: while both exduring and enduring objects are multi-q-located, only the former (again, barring some exotic cases noted below) have diachronic parts at every region at which they are multi-q-located. Indeed, clause (ii) of (D2.12) generally prevents an enduring object from having a diachronic part at any achronal slice of its path.

(D2.12)–(D2.14) thus delineate the important contrasts among the three modes of persistence. At the same time, it should be emphasized that these definitions are not watertight and are not intended to satisfy everyone.[49] Consider an enduring lump of clay that becomes a statue for only an instant (Sider 2001: 64–5). On (D2.7), the statue is a diachronic part of the lump at that instant. But endurance is widely regarded as being incompatible with the existence of diachronic parts. Also consider an organism composed of perduring cells and stipulate that the cells and their diachronic parts are the *only* proper parts of the organism (Merricks 1999: 431). By clause (iii) of (D2.13), the organism itself does not perdure.

Another exotic case[50] includes an object satisfying (D2.12) (hence, a bona fide endurer) but having a set of "finitely extended diachronic parts." It is unclear whether such an object could be regarded as enduring.

---

[48] This does not imply that exduring objects have *only* improper diachronic parts. It depends on how proper diachronic parthood at R is defined—the issue already considered above. In any case, the stage theorist should, of course, deny that an exduring object o is strictly identical with its t_1-stage, p_1, as well as with its distinct t_2-stage, p_2. If so, then under the aforementioned definition (D2.10′) of R-relativized proper diachronic parthood:

(D2.10′) $p_\parallel$ is a *proper diachronic part* of o at achronal region R $=_{df}$ (i) $p_\parallel$ is a diachronic part of o at R, (ii) $p_\parallel \neq o$,

at least one of p_1 and p_2 is a proper diachronic part of o (at t_1- or t_2-slice of o's path). On the other hand, if proper parthood at R is defined as *asymmetrical parthood* at R:

(D2.10) $p_\parallel$ is a *proper diachronic part* of o at achronal region R $=_{df}$ (i) $p_\parallel$ is a diachronic part of o at R, (ii) o is not a diachronic part of $p_\parallel$ at R,

then both p_1 and p_2 are improper parts of o, at different t-slices of its path. This, of course, does not entail that $p_1 = p_2$.

[49] It is unclear that such desiderata are met by any set of definitions offered in the literature on persistence.

[50] Suggested by a referee.

Relatedly, there could be an object satisfying clauses (i) and (ii) of (D2.13) but having only finitely extended proper diachronic parts. On (D2.13), such an object does not perdure, another intuitively wrong result. To handle possibilities of this sort, one would need to make full use of the appropriately defined notion of a "diachronically extended diachronic part," which lies outside the scope of this project.[51] Fortunately, cases of this sort are too remote to bear on the agenda of this book and we can safely ignore them. For our purposes, (D2.12)–(D2.14) provide good working accounts of the three modes of persistence.

We end this section with two more brief comments.

2.4.7. More on 'wholly present'

The above approach takes the notion of exact location as primitive. It aims to be neutral among the different modes of persistence (thus the sense in which a perduring object is exactly located at a 4D region of spacetime is the same in which an enduring object is exactly located at a 3D region) and to do justice to the idea of being "wholly present" at a moment of time, which has figured prominently in earlier discussions of endurance. Could the application of the relevant concept *being wholly present at R* to spatially composite objects be reduced to more basic mereological notions? The question has been thoroughly debated (see note 36), but that debate is rather peripheral to our concerns here. One reason to be skeptical about the prospect of such reduction is the dilemma of triviality and falsity often associated with it. If '*o* is wholly present at R' means that no part of *o* is absent from R, then one wants to know more about the notion of parthood at work. If parthood is taken to be a basic two-place relation (as it is in classical mereology) then the requirement that no part of *o* be absent from any region at which *o* is wholly present is incompatible with persistence through mereological change. On the other hand, if parthood is relativized to time (or more generally, to instantaneous subregions of spacetime, as in the present approach) then the requirement becomes trivial and thus empty.[52]

2.4.8. Should parthood be relativized? Could it be relativized?

One might argue that the basic two-place relation of parthood is as constitutive of mereology as the corresponding set membership relation

---

[51] See also n. 45.   [52] For an attempt to get around this problem, see Crisp and Smith 2005.

is constitutive of set theory and must, therefore, be protected. But many writers tend to think that the theoretical and heuristic benefits of relativized parthood outweigh its costs and that the notion is indispensable.[53] Indeed, much of "folk mereology" is temporally modified. My computer has just acquired a new video card, a part that it did not have before. To be sure, temporal modification of parthood is more familiar and transparent than parthood's relativization to a region of spacetime. But theories need to turn abstract at some point. Furthermore, a similar sort of relativization is needed by most accounts of persistence to explain change over time,[54] so treating parthood in the same way appears to be natural.

Another intriguing question is the *adicity* of the relativized parthood relation. The present work assumes that this relation is fundamentally *three*-place: x is a part of y at z, where z is a place holder for an achronal region of spacetime. This assumption looks natural[55] and appears to accommodate the relevant intuitions about multilocation and other important notions that figure in the debate about persistence. But one may have doubts[56] as to whether a *single* regional modifier can successfully relativize the instantiation of a parthood relation. Gilmore (2009a) has recently argued that three-place parthood confronts a number of problems, and that the best way to think of relativized parthood is in terms of a *four*-place relation: x at w is a part of y at z. The argument is extended and requires detailed consideration, which cannot be afforded here.

2.5. An Alternative Classification

The threefold classification of the views of persistence in a generic spacetime context developed above suits the project of this book quite well. It should be noted, however, that it is by no means the only possible such classification and may not even be the best one (if the notion of the best classification makes sense in abstraction from context). Below I briefly examine another attempt to categorize the views of persistence in a spacetime framework,

---

[53] "Relativizers" include Hudson 2001; Sider 2001; Bittner and Donnelly 2004; Crisp and Smith 2005; Balashov 2008; and Donnelly 2009.

[54] See §2.3 above and §§4.1 and 5.1 below.

[55] Especially in the context of the discussion in Chapters 4–8 of this book, where a moment of time (in a frame) may go proxy for a relevant achronal region of spacetime.

[56] Shared, among others, by a referee.

which is due to Cody Gilmore (2006, 2008) and motivated by quite different considerations.[57]

Gilmore begins by noting that there are two separate distinctions, which are often conflated in discussions of persistence: (i) the distinction between persisting by having temporal or diachronic parts and persisting without having such parts and (ii) the distinction between persisting by being temporally or diachronically extended and persisting without being extended in this way. An object can be said to persist by being diachronically extended iff it is located only at its path, and without being diachronically extended iff it is located only at achronal slices of its path:

(DG1) o persists by being diachronically extended $=_{df}$ (i) o persists, (ii) o is located only at its path o.

(DG2) o persists without being diachronically extended $=_{df}$ (i) o persists, (ii) o is located only at achronal slices of its path o.

This is a familiar distinction between single location and multilocation in spacetime.[58] Being singly located at an achronal region, however, does not entail having diachronic parts just as being multilocated does not entail lacking them, on any acceptable definition of the notion of 'diachronic part,' for example, on our earlier definition (D2.7).[59] Having a diachronic part is, therefore, logically independent from being diachronically extended, which creates logical space for four distinct views of persistence. To use Gilmore's terminology (2008: §4), a persisting object o:

(i) *locationally* and *mereologically endures* iff it persists by being located only at achronal slices of its path and without having any diachronic parts;

---

[57] I have put some of Gilmore's definitions in my own terms and introduced, along the way, some simplifications. One important distinction should also be noted: Gilmore's original classification provides room for "diachronically extended diachronic parts," whereas my rendition of it ignores them (see the note at the end of §2.4.6 above). I hope these differences do not change the spirit of Gilmore's approach, thus allowing me to illustrate the basic principles of his account in my own terms and in a framework that is friendly to my strategy. I chose to consider Gilmore's approach because its agenda comes closest to my own. Other classifications of the modes of persistence in spacetime have recently been developed by Rea 1998; Crisp and Smith 2005; Hudson 2006; Sattig 2006; and Parsons 2007.

[58] As Gilmore notes, the distinction defined by (DG1) and (DG2) is not exhaustive; it gives room for objects that are located at multiple finitely extended non-achronal regions but at no achronal regions. Such objects fit neither (DG1) nor (DG2). We also abstract from the issue of "o-eligibility" here. See §2.4.5.

[59] Gilmore uses a somewhat different but equally acceptable definition of 'temporal part.'

(ii) *locationally endures* but *mereologically perdures* iff it persists by being located only at achronal slices of its path and by having at least one diachronic part;

(iii) *locationally perdures* but *mereologically endures* iff it persists by being located only at its path and without having any diachronic parts;

(iv) *locationally* and *mereologically perdures* iff it persists by being located only at its path and by having at least one diachronic part.

Objects falling under (i) or (iv) are familiar enduring and perduring objects of older unsophisticated classifications. Such classifications, however, presuppose that multilocation at achronal regions precludes diachronic "segmentation" and that single location at a non-achronal region requires it. While locationally enduring but mereologically perduring objects and locationally perduring but mereologically enduring objects are rather *récherchés* they are logically possible. Indeed, their possibility is motivated by independent considerations, some of which have been already mentioned. A lump of clay made into a statue for only an instant (Sider 2001: 64–5) counts as a locationally enduring but mereologically perduring object. And a diachronically extended simple (or, more generally, any object that is diachronically extended while being at the same time "diachronically simple") counts as a locationally perduring but mereologically enduring object.[60] While the first case may be an extreme case suggested by reflection on the problem of coincident entities, the second is an increasingly popular example of an object that, despite its extension (in a relevant dimension), has no proper parts (in that dimension).

Given the importance of both cases in contemporary metaphysical debates it may be desirable to have a place for them in a classification of the competing views of persistence, and Gilmore's approach meets this desideratum in a very natural way. The present inquiry, however, abstracts from extended simples (see §1.2.1) and can also safely ignore instantaneous statues and their likes. On the other hand, I wish to regard exdurance as a bona fide mode of persistence, something that Gilmore's approach does not allow.[61] For my purposes, therefore, Gilmore's fourfold distinction would

---

[60] On achronally and diachronically extended simples see, e.g. Parsons 2000; Simons 2004; Hudson 2006: ch. 4; and McDaniel 2007.

[61] On Gilmore's approach, exduring objects do not persist. But I do not see why his definitions (DG1) and (DG2) could not be modified along the lines of my approach, by replacing in them 'location' with 'q-location':

be both too wide and too narrow. I shall henceforth adhere to my own threefold classification.

The spacetime framework underlying the foregoing discussion was kept neutral between various substantive views about the geometry of spacetime, such as classical physics and special relativity. The distinction between classical and relativistic spacetime, however, is crucial to the development of the ideas of this book. It is therefore time to introduce the basics of this distinction. It will be convenient to do it in two steps corresponding, roughly and broadly, to the historical sequence. This will also allow us to discuss new philosophical notions emerging from that context. These notions will play a role later. The reader familiar with the contemporary approach to spacetime theories can skip the following chapter without harm.

(DG1′) o persists by being diachronically extended $=_{df}$ (i) o persists, (ii) o is q-located only at its path o.

(DG2′) o persists without being diachronically extended $=_{df}$ (i) o persists, (ii) o is q-located only at achronal slices of its path o.

and thus be made exdurance-friendly. As already mentioned (n. 38), Gilmore himself makes an endnote provision for exdurance (Gilmore 2006: 230, n. 21) by suggesting an appropriate modification to the notion of 'path.'

3

Classical and Relativistic Spacetime

In 1908, three years after Einstein's famous paper introducing the theory of special relativity (SR), the German mathematician Hermann Minkowski showed that SR could be formulated as a theory about the geometry of the four-dimensional spacetime manifold. This had significant impact on subsequent developments leading all the way to general relativity and also greatly facilitated the understanding of the theory. Most textbook presentations of SR have since included the formalism of Minkowski diagrams, which is easy to learn and very instrumental in demystifying the counterintuitive results of relativity. In Minkowski formalism, a point in spacetime represents an *event*, an idealized occurrence in the physical world having extension in neither space nor time. Ordinary physical objects are represented by worldlines or full-blown four-dimensional worldworms (or, if you prefer, worldtubes), a collision of two objects is represented by the intersection of their worldlines (worldworms), and so on. Unlike the familiar three-dimensional world of experience, which undergoes change with time, the Minkowski world appears to be "static": nothing is happening there. Instead of being an external parameter of change, time is incorporated in the manifold itself as one of its dimensions, on a par with space. Minkowski drew the following lessons from his approach:

We should then have in the world no longer space, but an infinite number of spaces, analogously as there are in three-dimensional space an infinite number of planes. Three-dimensional geometry becomes a chapter in four-dimensional physics. . . . Henceforth space by itself, and time by itself, are doomed to fade away into mere shadows, and only a kind of union of the two will preserve an independent reality. (Minkowski [1908] 1952: 79–80, 75)

These statements have been widely quoted since but sometimes misunderstood. One should note, first, that the unification of space and time in a single 4D manifold does not, by itself, signify a distinctly relativistic result. Any physical theory, not just SR, can be formulated in a 4D spacetime framework. Classical mechanics is a case in point, as we shall see shortly. Where SR differs from it is in the different intrinsic geometrical *structure* it imposes on the spacetime manifold. The concept of such a structure is central to understanding SR as a theory about the geometry of spacetime. It is also responsible for many philosophical lessons that have been drawn from this theory. For example, the claim that SR is inconsistent with presentism, the view that only the present exists, is supported by the fact that the intrinsic geometry of Minkowski spacetime does not have enough structure to bring out the objective (that is, invariant or frame-independent) notion of the present.[1]

Secondly, having appreciated the elegance of spacetime diagrams and, in particular, the idea of representing physical objects by worldlines or worldworms (there is, again, nothing distinctively relativistic in the idea of such a representation), one can be easily tempted to simply *identify* objects with their worldlines (worldworms). Such identification would prejudge the issue between endurantism and perdurantism in a rather unfair way. As we have seen in Chapter 2, there is an unproblematic sense in which an enduring (or exduring) object has a four-dimensional path in spacetime without being itself four-dimensional. This suffices to show that eternalism does not entail perdurance. The upshot is that the language of spacetime theories is neutral with respect to the persistence controversy. The common basis of all such theories, classical as well as relativistic, is the 4D manifold of spacetime points (or point events).

Spacetime points should be thought of as being individuated, not by ascribing coordinates, or even geometrical relations, to them, but in some more rudimentary way—perhaps, by associating them with (actual and possible) idealized, infinitely short and small physical occurrences. Geometrization and coordinatization come much later in the process, and coordinatization never has a fully objective meaning even then. Before particular geometrical relations, let alone a particular coordinatization, are

---

[1] See §1.1.1–1.1.2 above on presentism versus eternalism and §3.2 below on the relationship between invariance and objectivity.

imposed on the manifold of spacetime points, they must already have identity.

Different spacetime theories are then distinguished in terms of the intrinsic structure they impose on this rudimentary basis. But what does it mean to attribute an intrinsic structure to something as intangible as spacetime? There are various ways to introduce the concept of such a structure. The simplest way is to ask what statements about events and objects a particular spacetime structure renders *meaningful*.[2] The more structure spacetime has, the more such statements become meaningful. Our eventual goal is to present Minkowski spacetime of SR and highlight those of its features that will play important roles in the subsequent chapters. But it will be convenient to start with a very different and much richer structure, that of Newtonian spacetime, and then effect a transition to the Galilean, or neo-Newtonian, framework by taking some of that structure away. This will teach us some important lessons, which can then be more easily applied to (what is often considered as) the more difficult transition from the Galilean (neo-Newtonian) to special relativistic spacetime. The reader familiar with these fundamentals can safely skip the rest of this chapter.

3.1. Newtonian Spacetime

The intrinsic structure of Newtonian spacetime can be informally described as follows (Huggett 1999: 192):

(Newtonian Spacetime):

Between any two spacetime points p and q, there is a definite Euclidean spatial distance $\Delta r(p,q) \equiv [(x_2-x_1)^2+(y_2-y_1)^2+(z_2-z_1)^2]^{1/2}$ and a definite temporal interval $\Delta t(p,q) \equiv t_2 - t_1$.

Here (x_1, y_1, z_1, t_1) and (x_2, y_2, z_2, t_2) are the coordinates of p and q in a Cartesian coordinate system associated with an inertial frame of reference.

To visualize this structure, imagine that the spacetime manifold is equipped with a complete set of parallel straight "position lines" and

---

[2] Our simplified exposition of spacetime theories follows the informal approach of Robert Geroch (1978) (although his presentation is tailored, eventually, to *general* not special relativistic spacetime). The definitions of Newtonian and neo-Newtonian spacetime below are adapted from Huggett 1999: ch. 10. For rigorous treatments, see Friedman 1983; Torretti 1983; and Earman 1989.

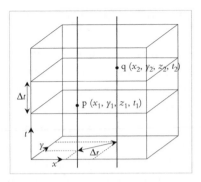

Figure 3.1. Spatial distance and temporal interval between two events p (with coordinates x_1, y_1, z_1 and t_1) and q (with coordinates x_2, y_2, z_2 and t_2) in Newtonian spacetime.

a complete set of parallel achronal three-dimensional Euclidean "time planes" (more precisely, time *hyperplanes*). The significance of position lines and time hyperplanes in Newtonian spacetime lies in the fact that they represent, in an important and deep sense, *positions in space* and *moments of time*. Any two pointlike physical events that occupy spacetime points on a given position line occur at the *same place*. Any two such events lying on the same time hyperplane occur at the *same time*.[3] The presence of this structure—the complete family of position lines and time hyperplanes—makes it clear what is involved in attributing two numbers, Δr and Δt, to an arbitrary pair of point events p and q. The spatial distance Δr between them is the Euclidean distance between the points at which the position lines containing p and q intersect any time hyperplane. Similarly, the temporal interval Δt between p and q is the temporal distance between the two time hyperplanes which host them.

Note that Δr has nothing to do with the "distance" between p and q in the four-dimensional spacetime. Note also that Figure 3.1 suppresses one of the dimensions of space (it is difficult to draw such diagrams in four dimensions). We shall continue using this simplification below by allowing ourselves the liberty to ignore one or even two spatial dimensions.

The spacetime structure just described derives its name from the fact that it represents, in modern terms, some essential elements of Newton's *metaphysics* of space and time, as outlined in the Scholium to Book I of the

---

[3] Cf. our earlier definition of a moment of time as a maximal achronal region of spacetime (§2.4.2).

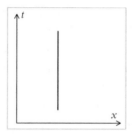

Figure 3.2. Particle at rest in Newtonian spacetime.

Principia. In particular, Newtonian spacetime supports the notion of *absolute rest* and, as a consequence, an absolute distinction between rest and motion, by making statements about the *velocity* of a given object meaningful. To see this, consider a small object o and its worldline in Newtonian spacetime. For any two points on the worldline, the spatial distance and temporal interval between them are determinate numbers Δr and Δt, which give all the resources needed to attribute to o a definite (instantaneous) velocity at any point at which it is located.[4] In particular, if Δr is zero throughout some finite interval of time $[t_1,t_2]$ then o is at (absolute) rest during that interval.

This illustrates the sense in which the spacetime structure described above provides a natural habitat for Newton's metaphysics of space and time. Contrary to Newton's own views, however, Newtonian spacetime is *not* adequate for Newtonian *physics*. The reason is that the structure imparts meaning to some statements that are *not allowed* to be meaningful by the laws of nature.

The laws of classical physics do not objectively privilege any inertial frame of reference. As epitomized in the Galilean principle of relativity, mechanical phenomena proceed in the same way in all inertial frames. Newton's laws of mechanics incorporate Galilean relativity by disallowing one to distinguish between a state of rest and a state of uniform motion with constant velocity: velocities do not figure in these laws, only accelerations do.

In modern terms, mechanical phenomena are said to be Galilean-invariant while the equations expressing their laws are Galilean-covariant.

---

[4] Or q-located. See §2.4. We shall ignore the distinction between location and q-location in this chapter.

Newton's first law says that a body unaffected by force retains its state of motion with constant velocity (including zero velocity, i.e. rest), while the second law says that the acceleration of a body is proportional to the net force on it: $\mathbf{F} = \mathbf{ma}$. Both laws retain their form under Galilean coordinate transformations: $\mathbf{r}' = \mathbf{r} - \mathbf{u}t$, $t' = t$.[5] The reason is that while the velocities of all objects are uniformly shifted by the same vector quantity $\mathbf{u}$ depending on the coordinate system in which they are measured, the objects' accelerations do not change. Consequently, accelerations but not velocities possess an objective status and must be backed by the intrinsic structure of spacetime.

To put it differently, spacetime must have enough structure to make statements of the relevant physical laws meaningful (in the case of Newtonian mechanics, such are statements attributing particular accelerations to objects, including zero acceleration). At the same time, spacetime must be devoid of any excessive structure that would make meaningful those statements that are not allowed to have meaning by the laws (in the present case, such are statements attributing particular velocities to objects, including zero velocity representing the state of rest). But contrary to this prescription, Newtonian spacetime geometry gives objective significance to velocities, not only accelerations. To eliminate the discrepancy between the structure of the laws and the geometrical structure of spacetime, the significance of velocities must be *abolished*; and the way to do it is to *deprive* Newtonian spacetime of its surplus structure.

Before we see how this program can be implemented it may be worthwhile to pause and reflect on why it is important to maintain a precise match between the intrinsic structure of spacetime, this universal arena of physical phenomena and processes, and the form of physical laws. It appears fairly obvious that spacetime must have at least *as much* structure as is necessary to make the statements of the laws meaningful. If a certain law attributes, say, a definite value of acceleration to a physical object, then this attribution had better make sense. Behind this principle "lies the realization that laws of motion cannot be written on thin air alone but require the support of various spacetime structures" (Earman 1989: 46). But why cannot spacetime have *more* structure than is needed to "support" the laws of physics? As noted above, such a surplus structure

[5] Here $\mathbf{u}$ is the constant velocity of the coordinate system (x',y',z',t') relative to (x,y,z,t).

would ascribe meaning to certain statements about events and objects that are not permitted to have meaning by the laws. But why should physical laws be thus empowered to decide which statements about events and objects in spacetime ought to be pronounced meaningless or, what comes to the same thing, what relationships among these entities ought to be prohibited from holding?

This issue is more controversial and it should be noted that the claim of physics to rule in this way over the geometry of spacetime could, in principle, be resisted. The history of modern physics presents outstanding cases of such resistance.[6] These cases show that introducing physically "idle" elements into the geometrical fabric of the world comes at a substantial cost of postulating a great deal of *conspiracy* among different quarters of physics, which is needed to ensure that the idle elements remain hidden in all physical situations. This is needed precisely because the physically idle elements of the surplus spacetime structure are not brought out by the laws. One can argue (see Janssen 2002*a*) that explanations based on such surplus structure are vastly inferior to those that take the more restricted structure as a *common origin* of all the diverse physical phenomena, in the same sense in which Ptolemaic "explanations" of planetary motions are inferior to Copernican explanations and special explanations of the origin of biological kinds are inferior to Darwinian explanations. Projects of this sort are therefore problematic and pursuing them requires really strong reasons. What could possibly count as a sufficiently strong reason in the context of spacetime theories? This question deserves more attention than can be afforded here. In what follows we shall presuppose[7] that, in the context of spacetime theories to be examined here, no good reason to oppose the usual strategy is available and, thus, the match between the geometry of spacetime and the form of physical laws must be preserved.

Our next task then is to perform a transition from the excessively rich Newtonian spacetime framework to a more flexible structure of *neo-Newtonian*, or *Galilean*[8] spacetime appropriate for classical physics.

---

[6] The most famous being the "Lorentz–Einstein controversy"; see Janssen 2002*b* for an excellent brief account of it.

[7] In conformity with our initial assumptions; see §1.1.2.

[8] These terms are widely considered to be synonymous, and for stylistic reasons we shall use both below.

3.2. Neo–Newtonian (Galilean) Spacetime

An important feature of Newtonian spacetime is that it has little room for *perspectivalism*: for drawing a distinction between the way things are in and of themselves and the way they appear in a certain perspective. The only legitimate perspective allowed by Newtonian spacetime is rigidly incorporated into its own intrinsic structure.

For example, the statement that a given material object is at rest should be understood as a *complete* statement expressing an *objective fact* about the world. The statement that two firecrackers explode one after another *at the same place* should similarly be interpreted as expressing an objective relation between the two events. And this appears to be quite wrong, both from the point of view of common sense and from the point of view of classical physics.

To illustrate, consider an observer sitting on object o_1 (Figure 3.3(a)). Such an observer sees a Newtonian world around him, in which he is at rest and object o_2 is in uniform motion. This state of affairs finds its geometrical manifestation in the fact that the worldline of o_1 is vertical while that of o_2 is inclined.

But the observer associated with o_2 will disagree with this description of the situation. She sees herself at rest (and draws her worldline vertically, as in Figure 3.3(b)) and o_1 as moving at a constant velocity toward her. And, of course, there are or could be numerous other objects in spacetime that are variously moving with respect to each other, and real or imaginary observers associated with them who see the situation in still other ways. Indeed, every observer would claim that he is at rest (and his worldline vertical) at the expense of others.

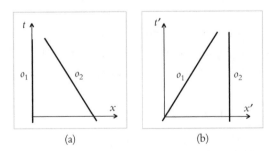

(a)   (b)

Figure 3.3. Two objects in relative motion.

Galilean, or neo-Newtonian, spacetime can be usefully introduced as a device "to stop this continual bickering among different [Newtonian] observers."[9] We simply declare that there is *no fact of the matter* as to which object is at rest and which are in motion. Consequently, there is no fact of the matter as to whose worldline is "vertical." Indeed, the notion of "verticality" cannot be defined in neo-Newtonian spacetime. The facts underlying these claims can be described as follows (see, e.g., Huggett 1999: 194):

(Neo-Newtonian (Galilean) Spacetime):

(i) Between any two spacetime points p and q, there is a definite temporal interval $\Delta t(p,q)$.

(ii) If p and q are simultaneous ($\Delta t(p,q) = 0$) then there is a definite spatial distance between them, $\Delta r(p,q)$.

(iii) For any line in spacetime, there is a fact of the matter about whether it is straight or curved.

Returning to Figure 3.3, in neo-Newtonian spacetime there is no fact of the matter as to whether o_1 or o_2 is at rest and which of their worldlines is "vertical." But how could it be? After all, Figure 3.3(a) represents object o_1 as being at rest and its worldline as "vertical" (i.e., parallel to the time axis of the coordinate system). And Figure 3.3(b) similarly attributes these characteristics to o_2 and its worldline. This is where the notion of perspectivalism comes into play. Figures 3.3(a) and 3.3(b) should be looked upon as different perspectival representations of the *same* state of affairs in different (inertial) reference frames. There is something "out there"—namely, the four-dimensional manifold, a certain geometrical structure intrinsic to it, and the physical contents of the manifold (i.e., objects, processes, etc.). But there are different ways of looking at this structure and contents—ways restricted to positions and states of motion of particular objects (or observers, if you prefer). The distinction is crucial. The structure "out there" is objective and real precisely because it is *invariant* with respect to change of perspective, or reference frame. On the other hand, additional elements brought by individual perspectives—for example, the "verticality" of certain lines in spacetime diagrams—are relative, perspective-restricted features of particular ways of looking at the objective reality. Such features are not part of the invariant spacetime

---

[9] Paraphrased from Geroch 1978: 46.

structure itself. But there is a clear sense in which the structure stands behind its seemingly conflicting perspectival representations.

3.3. Reference Frames and Coordinate Systems

Let us pause to discuss some important ideas introduced above, beginning with the notion of a *frame of reference*. For our purposes, an inertial reference frame can be associated with a family of parallel straight lines representing the paths of real or possible pointlike objects that are at rest (in that frame):

(D3.1) F is an *inertial reference frame* $=_{df}$ F is a maximal set of parallel straight diachronic lines in spacetime.

A reference frame represents a common perspective of all pointlike objects that are at rest in that frame, on the entire spacetime and its materials contents.

Associated with every inertial reference frame F is a set of distinct coordinate systems: $\{(x^F, y^F, z^F, t^F)\}$.

(D3.2) (x^F, y^F, z^F, t^F) is a coordinate system *associated* with (or *adapted* to) inertial reference frame F $=_{df}$ (i) (x^F, y^F, z^F) is a (Cartesian) coordinate system in 3D space having a certain spacetime point O as its origin, (ii) the time axis t^F contains O and is a member of F.

While different coordinate systems $(x^{F_1}, y^{F_1}, z^{F_1}, t^{F_1})$ and $(x^{F_2}, y^{F_2}, z^{F_2}, t^{F_2})$ associated with the same reference frame F share the *orientation* of their time axes t^{F_1} and t^{F_2}, they generally have different *origins* O_1 and O_2 and different orientation of their *spatial* axes. It is useful to think of two distinct coordinate systems associated with the same reference frame as representing somewhat different perspectives of two objects mutually at rest but separated by a distance and differently oriented in the common 3D space.

3.4. Galilean Transformations in Spacetime

Going back to Figures 3.3(a) and 3.3(b), they represent quite different perspectives on the common Galilean spacetime, the perspectives associated

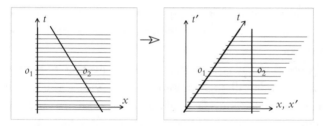

Figure 3.4. Galilean ("beveling-the-deck") transformation.

with different inertial reference frames. As already noted the translation between these perspectives is described by the Galilean coordinate transformations $\mathbf{r}' = \mathbf{r} - \mathbf{u}t$, $t' = t$. To make them more intuitive, let us follow Geroch (1978: 42 ff) and introduce a graphic model of switching between different "Newtonian perspectives" in spacetime. We shall imagine that spacetime is composed of a densely packed deck of cards representing time hyperplanes. The deck is pierced by straight "wires" representing the paths of inertial objects or, alternatively, positions in space in frames of reference in which such objects are at rest. The transformation whereby one goes from the perspective of Figure 3.3(a) to that of Figure 3.3(b) could then be effected by uniformly "beveling" the deck (see Figure 3.4).

Transformations of this sort induce different perspectives on a common geometrical structure. This common structure is none other than Galilean spacetime. As should be expected, it includes fewer intrinsic elements than Newtonian spacetime. Indeed, it comprises only those elements that survive, or remain invariant, under all "beveling-the-deck" transformations.

We already know what these invariant elements are (§3.2). They include, first of all, the entire family of time hyperplanes: beveling leaves them intact. This is not true of any single family of mutually parallel position lines. They get tipped under beveling and, consequently, what counts as the same position in space at two different times becomes a merely perspectival property of a pair of spacetime points (or events occurring at them) dependent on the state of "beveling." We say that there is no fact of the matter as to whether these events *really* occur at the same place, because Galilean spacetime does not have enough structure to bring this purported fact out.

We have already seen that facts about particular values of object velocities (or equivalently, about particular angles at which the objects' worldlines are inclined) also have no place in Galilean spacetime: the whole point of introducing this structure was to relativize states of rest and uniform rectilinear motion, in accordance with the Galilean principle of relativity (see Figure 3.4).

At the same time, it is important to reiterate that straight worldlines remain straight under all such transformations and, consequently, curved lines remain curved. Thus facts about the constancy and inconstancy (and furthermore, the rate of change) of objects' velocities—very unlike facts about their particular values—are invariant facts that are intrinsic to spacetime and do not depend on perspective. Upon reflection, this is precisely what is needed to secure a sought-for match between the geometry of spacetime and the form of the classical laws of motion. Recall: Newton's first law states that a body unaffected by force retains its state of rest or rectilinear motion with constant velocity. The law does not say anything about a particular *value* of that velocity (including zero value), but only about its constancy. This lawful state of affairs is perfectly accommodated in Galilean spacetime, which has no more, but also no less, intrinsic structure than is required to bring this fact out. The same reasoning applies to Newton's second law.

Galilean spacetime is therefore flexible up to arbitrary "beveling-the-deck" transformations. It endows certain worldlines, representing inertial non-accelerated motion, with the property of straightness (and curved worldlines, representing accelerated motion, with the invariant measure of curvature) but stops short of ascribing definite angles of inclination to them (that is, of ascribing definite velocities to objects they represent). What is the geometrical underpinning of this "middle ground?" Formally, it has to do with the affine structure of Galilean spacetime. Informally, one could imagine the deck of "cards" (i.e., the family of invariant time hyperplanes) being supplemented by a family of straight diachronic position lines—*all* of them—piercing the deck at all possible non-zero angles. This whole structure remains invariant under arbitrary "beveling-the-deck" transformations. Each subfamily of parallel position lines single out a particular perspective, or reference frame,[10] and can be used to impose a corresponding coordinatization on the common Galilean spacetime. In any such perspective, certain lines look "vertical" at the expense of others. But they are not *really* vertical. The necessity to draw them as vertical in any spacetime diagram is an

---

[10] Recall our earlier definition of this notion in §3.3.

artifact of graphic representation, not an intrinsic aspect of spacetime itself. If one wants to get a grip on what stands behind such representations, one should, in Geroch's apt expression (1978: 48), learn to "ignore verticality."

The fact that Galilean spacetime has less structure than its Newtonian predecessor translates into fewer statements about events and objects making sense. Indeed, only statements expressing the invariant features of spacetime—those unaffected by, or surviving under, the Galilean transformations—are meaningful. Here are examples of statements that do *not* make sense in Galilean spacetime (some of them have already occurred in our discussion):

- Non-simultaneous events p and q occur at the same place.
- Non-simultaneous events p and q are 7 meters apart.
- Object o's velocity is 5 m/sec.

Below are some examples of statements that *do* make sense in Galilean spacetime:

- Events p and q occur at the same time.
- Event p occurs 9 seconds later than q.
- Simultaneous events p and q are 7 meters apart.
- o's velocity is constant.
- o is accelerating.
- o_1 is moving at 5 m/sec relative to o_2.

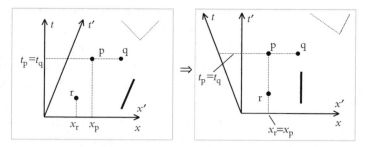

Figure 3.5. Features preserved under Galilean transformations: times of events (such as p, q and r), distances between simultaneous events (such as p and q), straight world-lines of particles, relative velocities. Such features are *invariant*, hence *objective*, as opposed, e.g., to distances between non-simultaneous events (such as p and r), absolute velocities of objects, and lightcones (the dotted lines in the upper right corner), which are relative features restricted to particular perspectives.

Figure 3.5 provides a summary of the invariant and non-invariant features of Galilean spacetime.

One moral of our discussion of Galilean spacetime is that what counts in it as *the same position in space at different times* is perspective-dependent and not part of the objective spacetime structure, whereas what counts as *the same moment of time at different places* is not so dependent.

3.5. Special Relativistic Spacetime

Galilean spacetime provides an ideal framework for the laws of classical mechanics: it has just enough structure to make the statements of such laws meaningful. But when the range of phenomena is expanded to include electromagnetism the Galilean framework becomes unable to accommodate them. The recognition of this fact eventually led to special relativity. To review briefly the logic of this transition, consider the following statement:

• The speed of light is 2.998×10^8 m/sec.

This statement attributes a certain velocity to a particular process. We know that such statements do not make sense in Galilean spacetime. Velocity is a frame-dependent quantity and thus does not remain invariant under "beveling-the-deck" transformations. Geometrically speaking, the worldlines of light (the dotted lines in Figure 3.5) change their angles of inclination in such transformations. Indeed, there is a perspective in which one of these worldlines becomes vertical. Light "stands still" in this perspective. But we know this is not what really happens. Not only does the above statement make sense, it is in fact *true*: the speed of light is the same in every reference frame and does not depend on the velocity of the light source.

One way to remedy the situation—the *right* way, as it turns out—is to modify the geometry of the spacetime manifold to enable the worldlines of light to stay intact under transformations between different perspectives associated with moving reference frames. The Galilean geometry does not allow this because "beveling the deck" uniformly tips all the worldlines embedded in spacetime, including the worldlines of light, in a "horizontal" (i.e. achronal) direction. But a cure can be purchased by excepting the worldlines of light, which (given a suitable choice of the units of length and time) are inclined at 45° to time hyperplanes, from being thus tipped.

Speaking informally, one needs to "make up" for their tipping-off by adjusting the orientation of *time hyperplanes* in a new reference frame. "Beveling-the-deck" transformations must be replaced by more complex "folding-unfolding" transformations in which *both* position lines and time hyperplanes get inclined by the same amount toward or away from the bisector of the angle between them. This transformation distorts the "shape" of spacetime in a more dramatic fashion than "beveling the deck": "folding-unfolding" causes certain portions of spacetime to be drawn closer to each other in one direction while stretching them away from each other in the other direction (see Figure 3.6). As a result, straight worldlines will in general be tipped one way or the other, but the worldlines of light, which are parallel to the bisector, will stay in place (see Figure 3.7).

The mathematical counterpart of the "folding-unfolding" procedure described above is, of course, the set of coordinate transformations distinctive of special relativistic spacetime and known as *Lorentz transformations*:

$$x'_p = \gamma(x_p - ut_p) \qquad v' = (v - u)/(1 - uv/c^2)$$
$$t'_p = \gamma(t_p - \frac{u}{c^2}x_p) \qquad \gamma \equiv (1 - v^2/c^2)^{-1/2}$$

Saving the constancy of the speed of light by making spacetime flexible with respect to these transformations comes at a price: when the orientation of the time hyperplanes is changed this affects the temporal relations between

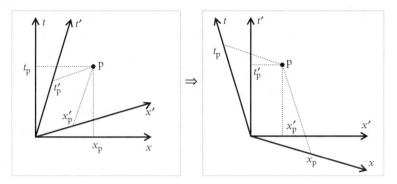

Figure 3.6. "Folding-unfolding" (Lorentz) transformation in special relativistic spacetime.

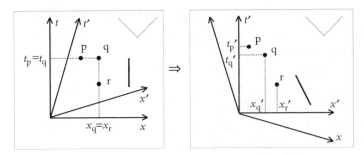

Figure 3.7. Events p and q are simultaneous in coordinate system (x,t): $t_p = t_q$ but not in (x',t'): $t'_p \neq t'_q$. Similarly, events q and r occur at the same position in (x,t): $x_q = x_r$ but not in (x',t'): $x'_q \neq x'_r$. Thus relativistic spacetime treats space and time truly on a par. The worldline of a material particle (the solid line in the diagrams) is tipped, hence the particle's velocity changes, under the transformation, but the worldlines of light (the dotted lines) stay intact, hence the velocity of light is preserved.

spatially separated events. Specifically, the facts about the *simultaneity* of such events do not survive under the "folding-unfolding" transformation. For example, events p and q, simultaneous in perspective (x,t) are no longer so in perspective (x',t') (see Figure 3.7).

Consequently, statements about simultaneity and, in general, about the temporal order of *spacelike separated* events[11] are meaningless in relativistic spacetime. There is no fact of the matter as to whether one of these events occurs earlier than, simultaneous with, or later than the other.

As we shall see momentarily, this highly counterintuitive result lies at the basis of the famous kinematic relativistic effects, such as length contraction and time dilation. Painful as it is, the relativity of simultaneity should be regarded as the culmination of the same "democratic" approach that had led earlier to Galilean relativity, and one's democratic sense should serve as a pain reliever. In the Galilean framework, spatial distance between events was relative to a reference frame and, in a particular frame, was a function of their temporal separation: space "depended" on time. But time did not "depend," in this sense, on space: temporal intervals between events were invariant and objective quantities. The relativistic framework eliminates this inequality of space and time by treating them more on a par. What counts

---

[11] In terms of our geometrical structure, two events are *spacelike separated* if the line connecting them is inclined at less than 45° to a time plane; if this angle is greater than 45°, events are said to be *timelike separated*. These notions will be defined more precisely in §3.7.

as the same moment of time at different spatial locations is no longer an objective feature of spacetime itself but a function of a particular spacetime perspective.

Another distinction of special relativistic spacetime from its Galilean predecessor should also be noted. In Galilean spacetime, the notion of *orthogonality* between position lines and time hyperplanes could not be defined: one had to "ignore the verticality" of certain position lines in frame-restricted representations. In special relativistic spacetime, on the other hand, this notion is well defined: position lines parallel to the time axis in a given frame are strictly orthogonal to the corresponding time hyperplanes, even though they do not always look that way. For example, in the right halves of Figures 3.6 and 3.7, the time axis t does not appear to be orthogonal to the time "hyperplane" x (shorthand for xyz; recall that two spatial dimensions are suppressed in those diagrams) but, in reality, it is. The concept of orthogonality at work here is grounded in the intrinsic geometry of special relativistic, or Minkowski, spacetime. The fact that some mutually orthogonal pairs of lines and hyperplanes do not look that way has to do with the non-Euclidean nature of this geometry, which defies visualization in our drawings. This inconvenience should simply be accepted as an unfortunate feature of graphic representation. Abstracting from it, one should regard both halves of Figures 3.6 and 3.7 as depicting the very same state of affairs. To paraphrase Geroch, in looking at a Minkowski spacetime diagram, one is supposed to ignore the fact that certain pairs of orthogonal lines and hyperplanes in spacetime do not look orthogonal.

3.6. Length Contraction and Time Dilation

Galilean ("beveling-the-deck") transformations preserved the spatial distances between simultaneous points and, hence, the spatial dimensions and shape of material objects. These transformations also preserved the temporal interval between spacetime points. Accordingly, the spatial dimensions and temporal intervals were intrinsic to spacetime and its inhabitants, not perspectival. The situation is drastically different in Minkowski spacetime.

Consider an idealized one-dimensional meter stick in special relativistic spacetime. Its path is represented as the shaded area in Figure 3.8. In the rest frame of the stick (x',t'),[12] its length (called *proper* length) is one meter. In the reference frame (x,t) uniformly moving in the direction of the stick, however, this length is contracted to 0.8 meters. This effect, known as Lorentz contraction, is purely relativistic in nature and has nothing to do with the mechanical contraction bodies may undergo under pressure. Relativistic contraction is a spacetime, not a dynamic phenomenon.[13] One way to explain it is to focus on what is involved in attributing length to an extended object, such as our stick, in a given perspective, or reference frame. Clearly, it involves taking the difference of the coordinates of the stick's ends in some coordinate system adapted to that frame—say, by using a similar measuring stick in the second frame. These coordinates must obviously refer to the *same* time. Put another way, the events of taking the measurements of these coordinates must be *simultaneous* and, hence, belong to the same time hyperplane in the reference frame under consideration. Geometrically, the sought-for length is just the length of the intersection of the stick's path with a time hyperplane in frame (x,t). Not surprisingly,

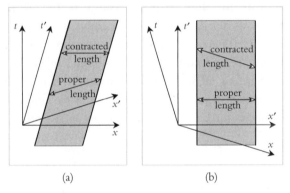

(a)   (b)

Figure 3.8. Lorentz contraction in special relativistic spacetime. (a) and (b) represent the same state of affairs. For convenience, both pairs of coordinate axes, (x,t) and (x',t'), are included in both diagrams. The lengths of the stick in the two frames are not drawn to scale.

---

[12] Strictly speaking, (x',t') is not a reference frame but one of the coordinate systems adapted to it. See §3.3 above. We shall henceforth allow ourselves to speak informally and refer to particular coordinate systems as "frames of reference," where context makes it clear which of the many systems associated with a given frame is under consideration. This manner of speaking is widely adopted in the literature.

[13] On the *standard* interpretation of SR, that is. For a different view of the nature of relativistic effects, see Brown 2005.

CLASSICAL AND RELATIVISTIC SPACETIME 59

it turns out to be different from the proper length of the stick.[14] One can see the relativity of simultaneity at play in fixing the reference of the frame-dependent notion of length.

Notably, length contraction and other kinematic relativistic effects are *reciprocal*. The length of a meter stick at rest can be measured by a meter stick in motion relative to it and found to be shorter than its proper length and than the length of the second stick (i.e. shorter than one meter). But the length of the second stick could similarly be measured by means of the first one and also found to be *shorter*; indeed, in the same proportion. But how could it be? It would appear that if x is shorter than y, then y must be *longer* than x.

Suppose the first meter stick is on the proverbial Einstein train.[15] Observers on the platform measure the length of the stick by marking the positions of its ends on their own meter stick, as the train zips in front of them at a considerable fraction of the speed of light, and find the distance between the marks to be 0.8 meters. There should definitely be a fact of the matter about the following: the right end of the moving stick must be flush with the right end of the stationary one when the latter is marked, while the left end of the moving stick must be in front of the 0.8-meter mark on the stationary one, at the moment that the mark is made (see Figure 3.9).

These facts must be intrinsic to the pair of sticks and taking the perspective of the train stick (in which it is at rest, while the platform stick is moving left) should not alter them. After all, one could imagine two arrays of small and quick creatures sitting on the sticks who manage to exchange super-rapid handshakes as they zip in front of each other. The fact that Sam, sitting on the right end of the train stick, shook hands with Susie, standing on the right end of the platform stick, at exactly the time she made a mark right in front of her, while Bob, riding the left end of the train stick, shook hands with Barbie, who happened to be standing on the 0.8-mark of the ground stick, at exactly the time that mark was made—*those* facts cannot be altered by simply changing the "angle" from which they are viewed in spacetime. But how could these facts fail to imply that, Susie and Barbie

---

[14] It turns out to be shorter than the latter, although it *looks* longer in Figure 3.8(b). This is an unfortunate distortion pertaining to the necessity to represent non-Euclidean relations inherent to relativistic spacetime in purely spatial Euclidean diagrams.

[15] The particular version of the Einstein's train *Gedankenexperiment* described below comes from Sartori 1996: 83–8.

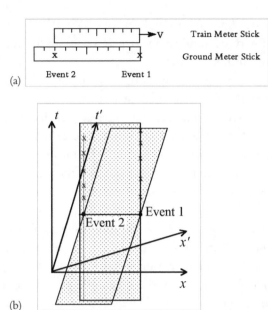

Figure 3.9. (a) The train meter stick is shorter in the ground frame, as evidenced by the marks made by ground observers on their stick. The snapshot of the mark-making is due to Sartori (1996: 86, fig. 15a) and is reproduced by permission: *Understanding Relativity: A Simplified Approach to Einstein's Theories*, by Leo Sartori, © 1996 by the Regents of the University of California. Published by the University of California Press. (b) A corresponding spacetime diagram shows the marks which, once made on the ground meter stick, "persist" over time.

having a rather tangible proof that their partners' train stick is shorter than their own, Sam and Bob must concede that Susie and Barbie's is *longer*?

Sam and Bob's stick is shorter than Susie and Barbie's in the latter's rest frame, while Susie and Barbie's stick is *shorter* than Sam and Bob's in *their* rest frame. The handshakes do not reveal this reciprocity simply because they are not suitable for measuring the length of moving objects in *both* frames. Indeed, the handshakes happen simultaneously in Susie and Barbie's common rest frame (i.e., ground frame) and, hence, can be used to measure the length of the train stick in that frame. But said handshakes do *not* happen simultaneously in Sam and Bob's rest frame. In fact, Sam touches Susie's hand *before* Bob touches Barbie's, in *that* frame. Figure 3.10 shows that this is consistent with the fact that the ground meter stick is *contracted* in the train frame. To detect this effect in a way similar to Susie and Barbie's detecting the contraction of the train stick, one would need to

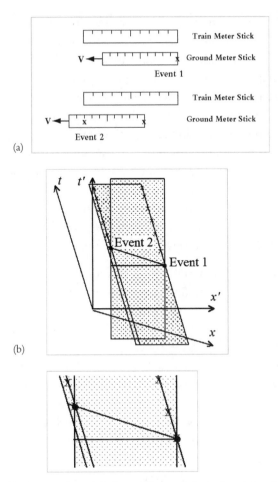

Figure 3.10. (a) Events 1 and 2 from Figure 3.9 depicted in the train frame. The snapshot comes from Sartori (1996: 86, fig. 15(b–c) and is reproduced by permission: *Understanding Relativity: A Simplified Approach to Einstein's Theories*, by Leo Sartori, © 1996 by the Regents of the University of California. Published by the University of California Press. (b) The corresponding spacetime diagram, with an area of interest blown up below.

take the coordinates of both ends of the ground stick at a certain moment of time in the *train* frame. This would require adding two more characters: Ted, sitting on the 0.8-mark of the train stick, and Tiffany, standing on the left end of the ground stick. Their handshake is simultaneous with Sam and Susie's handshake in the train frame and correctly shows the ground stick appropriately contracted.

Reciprocity of this sort can thus be traced to the relativity of simultaneity and is part and parcel of the *relativity principle*, one of the cornerstones of both classical and relativistic physics. The principle of relativity precludes one from privileging any (inertial) reference frame over any other. It does not make sense to ask which meter stick is *really* shorter; relativistic contraction is a relative not absolute phenomenon. The length of any object along a given spatial dimension is maximal in the object's rest frame (l_0, the *proper* length). When measured in a frame moving with velocity v relative to the first one along that dimension, it is always contracted by the factor $\gamma = (1-v^2/c^2)^{-1/2}$: $l = \gamma^{-1}l_0$. l is just as real as l_0; but both are relational not intrinsic properties, as they involve relations to the corresponding reference frames.

The same is true of *time dilation*, another kinematic relativistic phenomenon. We shall limit ourselves to listing the basic facts about it.

- The time elapsed between two events, or spacetime points, is *shortest* when measured in the (inertial) frame in which the events occur at the same place (*proper* time, Δt_0).
- In any other (inertial) frame, the time elapsed between the events is *longer* by the factor γ: $\Delta t = \gamma \Delta t_0$.
- In the (inertial) frame in which a certain clock is at rest, it measures proper time intervals between events or spacetime points that lie on its path.
- If that clock is monitored from an (inertial) frame moving with respect to that clock, it is found to run *slow* by factor γ in comparison with clocks that are stationary in the moving frame.
- This effect is perfectly reciprocal.

Figure 3.11 illustrates the effect of time dilation.

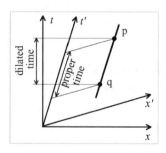

Figure 3.11. Relativistic time dilation.

3.7. Invariant Properties of Special Relativistic Spacetime

In view of the relativity of simultaneity, SR does not recognize any objective, perspective-independent notion of *being at the same time* extending across space, just as there was no objective notion of the sameness of position across time in the classical context. Being flexible to "folding–unfolding" transformations (also known as "Lorentzian boosts"), Minkowski spacetime cannot be uniquely foliated into a family of time hyperplanes. Rather every perspective on the spacetime reality induces its own foliation, each coupled with a corresponding family of position lines (see Figure 3.12).

We already know that special relativistic transformations preserve neither spatial distances nor temporal intervals between spacetime points. As a result, they also distort 3D spatial shapes of material objects embedded in spacetime: what is a sphere in one reference frame becomes an ellipsoid in another (see Figure 3.13).

But what are the invariant features of Minkowski spacetime, those that survive arbitrary "folding–unfolding" transformations? Some of them

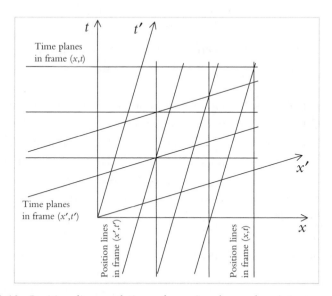

Figure 3.12. Position lines and time planes (i.e. hyperplanes) in two reference frames in Minkowski spacetime. Time in (x',t') "takes up" space in (x,t), and vice versa.

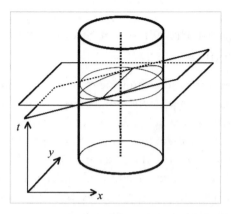

Figure 3.13. Two crisscrossing time-slices of a spherical object in Minkowski spacetime.

have already been noted. Although Lorentzian boosts in general change the orientation of diachronic lines (including position lines in a certain frame) and achronal hyperplanes (including time hyperplanes in a given frame), they preserve the straightness (as well as a certain well-defined measure of curvature) of the former and the flatness of the latter. They also maintain their mutual orthogonality (although this fact is not adequately represented in Euclidean diagrams). The transformations preserve the value of the speed of light (they were designed to do so) but not the speed of any material object, although they do preserve the constancy (and, hence, inconstancy) of it.

Finally, while special relativistic transformations change the spatial and temporal distances between spacetime points, Δr and Δt, they leave a certain *combination* of them intact. This is known as the *relativistic interval*: $I(\mathrm{p,q}) \equiv c^2 \Delta t^2 - \Delta r^2$. In many ways, the relativistic interval between spacetime points p and q, $I(\mathrm{p,q})$ is analogous to the spatial distance $d(\mathrm{A,B})$ between two points A and B in space (see Figure 3.14).

In a certain (profoundly mathematical) sense, the interval just *is* a (relativistic) distance between spacetime points. In that sense, Lorentzian boosts are similar to rigid rotations in space. Just as the latter do not affect distances between points in space, the former do not affect relativistic "distances" between spacetime points. But the analogy has its limitations. Indeed, the fact that time and space contribute opposite signs to the expression of the interval makes a lot of difference. Unlike distances in

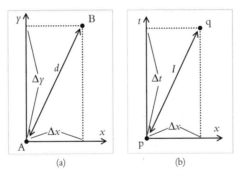

Figure 3.14. (a) Distance in space $d^2 = \Delta x^2 + \Delta y^2$ and (b) relativistic interval in spacetime $I = c^2 \Delta t^2 - \Delta x^2$.

space, relativistic intervals are not Euclidean, not positive-definite (may be negative), and may vanish for a pair of distinct points.

The sign of the relativistic interval forms the basis of an important threefold classification of invariant relations between spacetime points. This classification also has an intuitive geometrical manifestation, the famous lightcone structure of Minkowski spacetime.

(Lightlike Separation): Spacetime points p and q are lightlike separated $=_{df} I(p,q) = 0$.

p and q are lightlike separated provided that, during the time elapsed between them, Δt, light exactly covers the spatial distance between them, Δx: $c\Delta t = \Delta x$. In other words, p and q can be connected by a light signal. Clearly, if this holds in some inertial frame, it holds in each and every (see Figure 3.15):

(Timelike Separation): p and q are timelike separated $=_{df} I(p,q) > 0$.

p and q are timelike separated insofar as, during the time between them, light travels *farther* than the distance between the points: $c\Delta t > \Delta x$ (see Figure 3.16(a)).

This means that the points could be connected by some physical process propagating at a subluminal speed and, therefore, could belong to a path of a material object. If such an object is a *clock*, it will measure the *proper time* between p and q. From its perspective, p and q occur at the same place (Figure 3.16(b)) and the proper time elapsed between them is $\Delta t' = \sqrt{I}$. The positive interval thus provides a measure of proper time between timelike separated points.

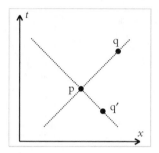

Figure 3.15. p and q are lightlike separated, and q is to the future of p. p and q′ are lightlike separated, q′ being to the past of p. $I = c^2\Delta t^2 - \Delta x^2 = 0$, hence $c\Delta t = \Delta x$. Light *just makes it* from p to q (and from q′ to p): during the elapsed time between the two events, Δt, light has just enough time to travel the distance between the events.

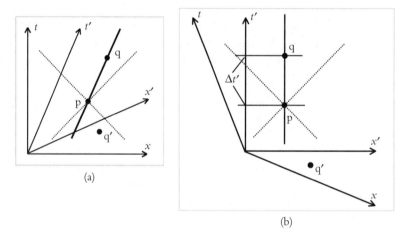

Figure 3.16. (a) p and q are timelike separated, and q is to the future of p. p and q′ are timelike separated, q′ being to the past of p. $I = c^2\Delta t^2 - \Delta x^2 > 0$, hence $c\Delta t > \Delta x$. During the elapsed time between the two events, Δt, light has enough time to travel farther than the distance between the events. If p and q are timelike separated there is a possible clock running through both p and q. According to this clock (i.e., in its rest frame), the spatial distance between p and q is zero, $\Delta x' = 0$ (see (b)), and the time elapsed between p and q is $\Delta t' = \sqrt{I}$.

(Spacelike Separation): p and q are spacelike separated $=_{\text{df}} I(\text{p,q}) < 0$.

p and q are spacelike separated when they are too far apart in space; light never makes it from p to q: $c\Delta t < \Delta x$ (Figure 3.17(a)).

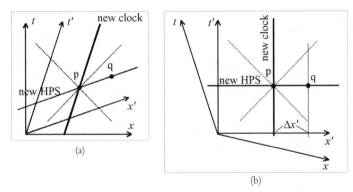

(a)

(b)

Figure 3.17. (a) p and q are spacelike separated. $I = c^2\Delta t^2 - \Delta x^2 < 0$, hence $c\Delta t < \Delta x$. p and q are too "far apart" in space; during the elapsed time between the two events, Δt, light does not have enough time to travel the distance between the events. There is *no* clock that could run through both events; but there is a frame in which the events are *simultaneous*, i.e., lie on the same hyperplane of simultaneity (HPS). See (b). According to a clock corresponding to that frame, the time elapsed between p and q is zero, $\Delta t' = 0$, and the spatial distance between them is $\Delta x' = c\sqrt{-I}$.

If p_1 and p_2 are spacelike related, there is no clock that could run through both; but there is a frame in which these points are *simultaneous*. In that frame, the spatial distance between them is $\Delta x' = c\sqrt{-I}$ (see Figure 3.17(b)). The negative interval thus provides a measure of spatial distance between spacelike separated points (in the frame in which they are simultaneous).

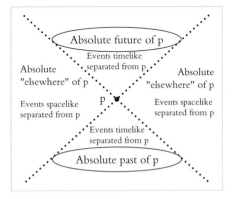

Figure 3.18. The lightcone structure of Minkowski spacetime.

Consider now an arbitrary point p. The set of spacetime points that are lightlike separated from p forms a three-dimensional lightcone centered on it (see Figure 3.18).

The lightcone divides the rest of spacetime into three separate regions: two almost unconnected regions (the only connection is the point p itself), the *absolute past* and *absolute future* of p, comprised of points that are timelike or lightlike separated from p, and a connected region of points that are spacelike separated from it, known as the *topological present* of p, or its "absolute elsewhere."

The significance of these regions lies in their invariance with respect to Lorentzian boosts, which do not change their membership (see Figure 3.19). Events in the absolute future of p occur later than p in *any* reference frame. Due to its absolute chronological precedence, a physical event occurring at p can *causally* influence events in its absolute future (e.g. a physical event at q^+). Similarly, physical events in the absolute past of p, such as an event located at q^-, chronologically precede p in the absolute sense and, therefore, the physical event located at p can be causally affected by them. The event at p can also causally influence and be influenced by events that are lightlike separated from p (e.g. an event located at q^0); such an influence must propagate with the maximum possible speed ($= c$).

On the other hand, no *objective* temporal order exists between p and points in its "absolute elsewhere," such as q^*. Since q^* is spacelike separated from p, there is a reference frame in which the event located at q^* occurs later than that located at p, a frame in which the former occurs earlier than

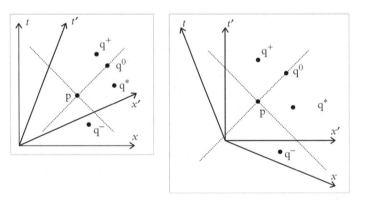

Figure 3.19. Objective causal and chronological relations in Minkowski spacetime.

the latter, and a frame in which they occur simultaneously. In the absence of a determinate chronological order between them, spacelike separated events cannot be causally related. This is good news, for in order to be so related, such events must be connectable by a physical process propagating faster than light, and there are no such processes in nature.[16]

To sum up, the intrinsic structure of Minkowski spacetime is chronologically and causally ordered; but the ordering is partial.[17] We end this section and chapter by providing an informal description of this structure.

(Special Relativistic Spacetime):

(i) For any two spacetime points p and q, there is a definite relativistic interval $I(p,q)$.

(ii) For any line in spacetime, there is a fact of the matter about whether it is straight or curved.

(iii) For any achronal surface in spacetime, there is a fact of the matter about whether it is flat or curved.

---

[16] Unless there are *tachyons*: hypothetical particles moving faster than light.

[17] The sense in which Minkowski spacetime is partially ordered is the sense in which its points can be ordered by the relation *being identical or timelike-or-lightlike separated from and to the future of*, which is reflexive, antisymmetric, and transitive.

4

Persisting Objects in Classical Spacetime

In Chapter 2 the rival doctrines of persistence were stated in a generic spacetime framework. Now we need to situate them more precisely in the context of particular spacetime structures. In this chapter we adapt the generic framework to classical (Galilean) spacetime. Given the background of Chapter 3, this task is relatively straightforward and recovers a familiar set of notions at work in earlier discussions of persistence. Although engaging those earlier considerations is not central to this book's agenda, in §§4.2–4.5 I review and identify a weak spot in a particularly influential argument for 4Dism, the argument from vagueness. Further implications of this criticism are then explored in the concluding section of this chapter.

4.1. Enduring, Perduring, and Exduring Objects in Galilean Spacetime

The relation of absolute chronological precedence ($<$) in Galilean spacetime (ST^G) coincides with the relation of absolute temporal precedence: $p_1 < p_2 \leftrightarrow t_1 < t_2$, where (x_1,y_1,z_1,t_1) and (x_2,y_2,z_2,t_2) are the coordinates of p_1 and p_2 in any Cartesian coordinate system associated with any inertial frame of reference. Accordingly, a region R of Galilean spacetime is achronal iff it is a subregion of an absolute time hyperplane. That is:

(D4.1^G) Region R of ST^G is *achronal* $=_{df} \forall p_1, p_2 \; (p_1, p_2 \in R \rightarrow t_1 = t_2)$.

And a moment of time (= a maximal achronal region) is simply a time hyperplane in ST^G:

(D4.2^G) R is a *moment of time* in ST^G =$_{df}$ R is a time hyperplane in ST^G.

A brief comment on the notation. The definitions of various persistence-related concepts in this section should, for the most part, be looked upon as *adaptations* of the generic definitions of §2.4 to the particular environment of Galilean spacetime. Making this explicit would require systematic replacement of all the predicates (such as *achronal, moment of time*, etc.) with their Galilean counterparts (e.g. *achronal-in-ST^G*, or *Galilean-achronal*, etc.), which would create needless complexity at almost every step. To avoid it, we adopt a convention whereby all the appropriate modifiers (i.e. 'Galilean,' in this chapter, and 'Minkowskian' in the next chapter) shall be understood as being supplied by context, the superscript '^{G}' (which will give way to '^{M}' in the next chapter) serving as an occasional reminder. This harmless simplification will allow us to focus on the really important issues.

In particular, the definitions of *q-location* (*quasi-location*) and *path* can be taken directly from Chapter 2, with a minimal change of notation:

(D4.3^G) *o* is (exactly) *q-located* at R in ST^G =$_{df}$ one of *o*'s (non-modal) counterparts is (exactly) located at R.[1]

(D4.4^G) Spacetime region *o* is the *path* of object *o* in ST^G =$_{df}$ *o* is the union of the spacetime region or regions at which *o* is *q-located*.

According to our older generic definition (D2.5), *o persists* just in case *o*'s path is non-achronal. Adapted to Galilean spacetime, this boils down to the requirement that *o*'s path intersect at least two distinct moments of time.

(D4.5^G) *o* persists in ST^G =$_{df}$ $\exists p_1, p_2 \in o$, $t_1 \neq t_2$.

As before (§2.4.4), we take a three-place relation '*p* is a part of *o* at achronal region R' as a primitive and use it in defining achronal and

---

[1] As before, those who are dissatisfied with the broad sense of 'counterpart' at work in (D4.3^G), may choose a less elegant equivalent of (D4.3^G):

(D4.3$^{G'}$) *o* is (exactly) *q-located* at region R of ST^G =$_{df}$ *o* is (exactly) located at R or one of *o*'s (non-modal) counterparts is (exactly) located at R.

diachronic parthood relations. But here further simplifications are in order. The earlier generic definitions of 'achronal part of o at achronal region R' (D2.6) and 'diachronic part of o at achronal region R' (D2.7) generalized the concepts of spatial part and instantaneous temporal part to the spacetime framework. In Galilean spacetime, however, all and only achronal regions are moments of absolute time. This effectively reduces some of the generic notions of Chapter 2 to their more familiar classical predecessors. In particular, an achronal slice $R_\perp$ of R in ST^G is simply the intersection of R with a moment of time:

(D4.8^G) $R_\perp$ is an *achronal slice* of R in ST^G $=_{df}$ $R_\perp$ is a non–empty intersection of a moment of time (i.e., a time hyperplane) with R.

Accordingly, I shall refer to the achronal slice of R at t in ST^G simply as the 't-slice of R' or '$R_{\perp t}$'. This brings the concepts of achronal and diachronic parthood at achronal region R closer to the older concepts of *temporal part at t* and *spatial part at t*. In what follows I shall sometimes use such simpler notions, where context makes it clear that 't' refers not to an entire hyperplane of absolute simultaneity but to a rather small subregion of it—$o_{\perp t}$:

(D4.6^G) $p_\perp$ is a *spatial part (s-part)* of o at $o_{\perp t}$ in ST^G $=_{df}$ $p_\perp$ is a part of o at $o_{\perp t}$.

(D4.7^G) $p_\parallel$ is a *temporal part (t-part)* of o at $o_{\perp t}$ in ST^G $=_{df}$ (i) $p_\parallel$ is located at $o_{\perp t}$ but only at $o_{\perp t}$, (ii) $p_\parallel$ is a part of o at $o_{\perp t}$, and (iii) $p_\parallel$ overlaps at $o_{\perp t}$ everything that is a part of o at $o_{\perp t}$.

Finally, we adapt the notion of *object-eligible achronal slice* of that object's path (see §2.4.5) to the geometry of Galilean spacetime:

(D4.11^G) $o_{\perp t}$ is an *o-eligible t-slice of o's path* o $=_{df}$ either o itself or o's t-part at $o_{\perp t}$ is q-located at $o_{\perp t}$.

Based on this stock of notions, endurance, perdurance, and exdurance in Galilean spacetime can be defined as follows:

(D4.12^G) o *endures* in ST^G $=_{df}$ (i) o persists, (ii) o is located at every o-eligible t-slice of its path, (iii) o is q-located only at o-eligible t-slices of its path.

(D4.13^G) *o perdures* in ST^G $=_{\text{df}}$ (i) *o* persists, (ii) *o* is q-located only at its path, (iii) the object located at any *o*-eligible *t*-slice of *o*'s path is a proper *t*-part of *o* at that slice.

(D4.14^G) *o exdures* in ST^G $=_{\text{df}}$ (i) *o* persists, (ii) *o* is located at exactly one region, which is an *o*-eligible *t*-slice of its path, (iii) *o* is q-located at every *o*-eligible *t*-slice of its path.

As noted in Chapter 2, these definitions are not watertight, but they bring out all the essential differences among the three modes of persistence in Galilean spacetime.[2]

Each mode of persistence comes with a characteristic *regional modification* scheme (or schemes). In the earlier informal discussion (§2.3) such schemes were looked upon as providing analyses of simple statements of the form '*o* has Φ at *t*.' Now the ascription of properties must be relativized to achronal regions of spacetime, namely, to achronal slices of *o*'s path. However, since in ST^G all the achronal regions of interest can (in ordinary cases) be indexed by moments of absolute time, we can, for the purpose of illustration, treat the simple temporal modifier as shorthand for the regional modifier.

The following is a brief summary of the analyses of regional modification in the competing views of persistence, beginning with endurance, which allows three somewhat different schemes:[3]

(D4.15^G) Enduring object *o* has Φ at *t* (i.e., at $o_{\perp t}$) in Galilean spacetime $=_{\text{df}}$ *o* bears Φ-at to *t*.

(D4.16^G) Enduring object *o* has Φ at *t* (i.e., at $o_{\perp t}$) in Galilean spacetime $=_{\text{df}}$ *o* has Φ-at-*t*.

(D4.17^G) Enduring object *o* has Φ at *t* (i.e., at $o_{\perp t}$) in Galilean spacetime $=_{\text{df}}$ *o* has$_t$ Φ.

Perdurance and exdurance, on the other hand, naturally go along with the following canonical accounts of temporal predication in Galilean spacetime:

---

[2] See §2.4.6 for a detailed explanation of these differences in the generic spacetime framework and a brief discussion of some exotic counterexamples. Such exotica emerge in the Galilean framework too; in particular, (D4.7^G) proclaims an instantaneous statue (see Sider 2001: 64–5) to be a *t*-part of an enduring lump of clay, and (D4.13^G) categorizes a "Merricks organism" (Merricks 1999: 431) as not perduring. Cody Gilmore warns me that counterexamples do not end there.

[3] D4.15^G describes (Galilean) relationalism (not to be confused with spacetime relationism), D4.16^G (Galilean) indexicalism, and D4.17^G (Galilean) adverbialism. See §2.3.1 for an informal discussion of these schemes.

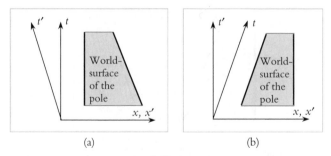

(a)   (b)

Figure 4.1. A contracting pole in two inertial reference frames.

(D4.18^G) Perduring object o has Φ at t (i.e., at $o_{\perp t}$) in Galilean spacetime
$=_{df}$ o's t-part has Φ.

(D4.19^G) Exduring object o has Φ at t (i.e., at $o_{\perp t}$) in Galilean spacetime
$=_{df}$ o's t-counterpart has Φ.

To illustrate these ideas further, consider a 2-meter-long folding pole in Galilean spacetime. At a certain moment, it starts to contract until its length is reduced to 1 meter. A Galilean spacetime diagram in Figure 4.1(a) represents the path of the pole in a reference frame in which its left end is at rest (as usual, we neglect two spatial dimensions). The diagram in Figure 4.1(b) represents the same states of affairs in another reference frame, the rest frame of the right end.

We are interested in the objective Galilean-invariant state of affairs itself, not in its various perspectival representations. Consequently, we need to make an intellectual effort and "ignore the verticality"[4] of certain world lines in Figures 4.1(a) and 4.1(b). In fact, we need to *identify* the two diagrams, by allowing ourselves freedom to perform arbitrary "deck-beveling" transformations.[5] The resulting invariant structural features are shown in Figure 4.2. They include the "slicing" of the pole's path by the family of absolute time hyperplanes and the lengths of the slices on each such plane. The choice of the zero time hyperplane is conventional.

Every time plane in Figure 4.2 features a determinate intrinsic property: the length possessed by the pole at the time in question. The next step is

---

[4] In Geroch's (1978: 48) apt expression.   [5] See §3.4.

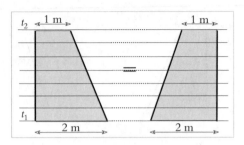

Figure 4.2. The folding pole in Galilean spacetime.

to provide an analysis of this interaction between time and attribution of properties.

For the endurantist, the pole is a 3D entity extended in space but not in time. It is located at t-slices of its path and any such slice features the full set of properties the pole has at a corresponding time, including its length. On relationalism, the pole comes to have the property of being 2 meters long at t_1 and 1 meter long at t_2 by bearing the relation *2-meters-long-at* to t_1 and *1-meter-long-at* to t_2.[6] On indexicalism, the pole accomplishes the same feat by exemplifying two time-indexed properties, *2-meters-long-at-t_1* and *1-meter-long-at-t_2*. On adverbialism, the pole possesses the simple property *2-meters-long* in the t_1-ly way, and another property, *1-meter-long*, in a different, t_2-ly way.

For the perdurantist, on the other hand, the pole is a 4D entity having extension both in space and time. It persists by having distinct momentary t-parts at t-slices through its path. When we say that the pole is 2 meters long at t_1 and 1 meter long at t_2, what we really mean is that the pole's t_1-part has the former property and its t_2-part the latter.

For the exdurantist, the pole is a 3D entity that is q–located at multiple t-slices through its path, thanks to having distinct t-counterparts at each such slice. The pole comes to be 2 meters long at t_1 and 1 meter long at t_2 by having a t_1-counterpart and a t_2-counterpart, which have these respective lengths *simpliciter*. (Remember that the t-counterpart relation is reflexive.)

We end this section by noting that o-eligibility in Galilean spacetime (D2.11^G) and the definition of the object's path (D4.4^G) trivially entail (Galilean) "Achronal Universalism":

---

[6] As noted above, in simple contexts 't_1' and 't_1' come in handy as useful shorthand for '$o_{\perp t_1}$' and '$o_{\perp t_2}$'.

(AU^G) (i) Any enduring object is located at *every* t-slice of its path (in Galilean spacetime), (ii) any perduring object has a t-part at every t-slice of its path, and (iii) any exduring object is q-located at every t-slice of its path.

Indeed, given that (i) a persisting object or any of its t-parts is located, or q-located (depending on whether o endures, perdures, or exdures) at all and only o-eligible t-slices of its path, and (ii) o's path is a union of regions at which it is (q-)located, it follows that every t-slice of o's path is automatically o-eligible. The reason, of course, is that Galilean spacetime excludes "crisscrossing" achronal slices (see §2.4.5), thus making the notion of (Galilean) o-eligibility, in effect, redundant. The modifier 'o-eligible' could well be eliminated from all three definitions (D4.12^G–D4.14^G) of the modes of persistence in Galilean spacetime. As one may expect, the situation will be interestingly different in Minkowski spacetime (§§5.5 and 5.6 below).

4.2. The Argument from Vagueness

The classical setting described above comes rather close to the context in which the issue of persistence had been widely discussed before the recent effort at reformulating it in the spacetime framework. In many ways, time can be regarded, in the Galilean setting, as being "separate" from space (see §§3.2–3.4). Accordingly, on the assumption that spacetime is classical, the debate can be conducted, to some extent, in older terms.

The "problem of temporary intrinsics" (§2.3.2) is one of the several issues that figured prominently in many earlier discussions of persistence, mostly in the context of defending 4Dism. Other popular arguments supporting 4Dism over 3Dism include the need to avoid co-located objects,[7] the need to allow for time travel and other exotica.[8] Influential arguments

---

[7] For details and references to earlier work, see Rea 1997; Sider 2001: ch. 5 and 2008: 247–57. See also McGrath 2007a, 2007b for critical discussion of co-location-inspired arguments for 4Dism.

[8] See Sider 2001: §4.7. However, Gilmore 2007 has argued that a certain type of time travel poses a problem for perdurantism. For objections to Gilmore's argument, see Eagle 2009a and 2009b. For a reply, see Gilmore 2009b.

defending 3Dism over 4Dism include the "modal inductility" objection to perdurantism[9] and the rotating disk scenario.[10] The literature on these and related topics is extensive,[11] but the focus of the present study is different and prevents us from doing justice to them. One argument, however,—the argument from vagueness developed most fully in Sider (2001: §4.9)—holds a special place in the debate about persistence, and broader implications of its rejection are relevant to our concerns.

The significance of the argument from vagueness is underscored by the fact that some authors take it to be central to the case for 4Dism.[12] The argument has provoked diverse responses questioning virtually every assumption that goes into its construction.[13] Below I add my voice to the critics. I hasten to note that although I share with them the belief that the argument fails because of its reliance on a controversial thesis about material composition known as *mereological universalism*, my recipe for dealing with that thesis and my resulting critical strategy are different.

Sider's argument includes three steps. Step 1 defends a version of David Lewis's argument (1986: 212 f) for unrestricted composition *at a time*, or *synchronic universalism*:

(SU) Any class of objects existing at *t* has a fusion at *t*.

Step 2 employs considerations parallel to those at play in Step 1 to argue for *diachronic universalism*—roughly, the view that, for any interval of time (which need not be continuous) and objects existing at various moments in it, there is something they compose over the interval:

---

[9] See van Inwagen 1990*a* for the original objection, and Heller 1993 and Sider 2001: 218–24 for some responses.

[10] The scenario, originally attributed to Kripke and Armstrong 1980, is debated in Zimmerman 1998*b*, 1999; Lewis 1999; Callender 2000*b*; Sider 2001: §6.5; Hawley 2001: ch. 3. Butterfield 2006*a* has recently invoked non-achronal temporal parts to rebut the rotating disk argument on behalf of the perdurantist.

[11] For excellent recent reviews, see McGrath 2007*b*; Hawley 2008; and Sider 2008.

[12] Thus Sider admits that it is "one of the most powerful" (2001: 120) and ranks it next only to the argument from "coincident entities." Ned Markosian takes it be "the most important and powerful argument in [Sider's] book" (2004*b*: 665). And Kathrin Koslicki claims that "in the end . . . everything turns on the [argument]. . . . If it were not for [it], there would be a relative stand-off between [3Dism and 4Dism], given the rest of Sider's evidence" (2003: 108). I suppose many 4Dists and some 3Dists would resist the claim that *everything* turns on the argument from vagueness. But clearly that argument carries a lot of weight in the persistence debate, as do other considerations related to vagueness. Cf. Hawley 2001: ch. 4.

[13] See, e.g. Koslicki 2003; Markosian 2004*b*; Nolan 2006.

(DU) For any set of objects $\{x(t)\}$, such that for any $t \in T$, $x(t)$ exists at
      t, there is an object y existing at all $t \in T$, but at no $t \notin T$, such
      that, at t, $y = x(t)$.[14]

Finally, Step 3 uses this result to show that temporal parts exist.

Other critics have tended to treat the first two steps as a package deal that
could and should be resisted as a whole.[15] In difference from them, I have
nothing to say about composition at a time. But I reject the assumption,
shared by Sider with most of his critics, that synchronic composition and
what Sider calls "minimal diachronic fusion" are sufficiently similar to use
considerations inspired by Step 1 (i.e. the argument for (SU)) to bear on
the latter. My objection to a crucial premise of Step 2 turns on the relevant
aspect of dissimilarity between these two cases.

If I am right then (a) the argument from vagueness is unsound and (b)
rejecting (DU) is sufficiently motivated. While the 3Dist may welcome
(a) as undermining the general case for 4Dism, I am inclined to consider
both (a) and (b) as encouraging important domestic restructuring in the
4Dist camp. Although many 4Dists are, in fact, diachronic universalists,
they need not and, I urge, should not be. If (DU) proves unhelpful in
supporting 4Dism, it becomes a burden. And liberating oneself of the
commitment to diachronic "trout-turkeys" should, I think, be perceived
as a relief, even if this, by itself, does not help with their notorious
synchronic ilk.

I begin by reviewing the argument and other criticisms of it. I offer my
own critique in §4.4 and discuss some objections and implications in §§4.5
and 4.6.

4.3. From Minimal D-Fusions to Temporal Parts

Step 1 of the argument from vagueness, which is a tightened-up version
of Lewis's earlier attack on restricted composition, employs some technical

---

[14] The notion of existing at a time and the relations of parthood and identity at a time are intended,
in the present context, to be neutral between 3D and 4D, to avoid prejudging any important issues.
For more details, see Sider 2001: §§3.2–3.3 and 4.9.2, and §§4.3–4.5 below.

[15] Considerations for rejecting the deal include noting that composition may be brutal (Markosian
2004b) and that Sider's support of the important premise that composition is never vague is dialectically
weak (Koslicki 2003).

notions.[16] A *case of composition* is a possible situation involving a class
of objects at a time, for which the question of whether they compose
anything, or have a fusion, can be raised. A *continuous series of cases* is
a finite series of cases of composition in which any two adjacent cases
are extremely similar in all respects that may be deemed relevant to
composition. A *sharp cutoff* in a continuous series is a pair of adjacent cases
differing in whether composition occurs.

The premises of Step 1 are as follows:

P1:   If not every class has a fusion, then there must be a pair of cases
      connected by a continuous series such that in one, composition
      occurs, but in the other, composition does not occur.

P2:   In no continuous series is there a sharp cutoff in whether composition
      occurs.

P3:   In any case of composition, either composition definitely occurs, or
      composition definitely does not occur.

Suppose composition (at a time) is restricted. Then there is a continuous
series connecting a case of composition to a case of non-composition (P1).
By P3, there must be a sharp cutoff in this series, which is prohibited by
P2. Therefore composition is unrestricted.

P1 is the least problematic of all the premises of Step 1.[17] Sider's
defense of P2 exploits the intuition that any sharp cutoff would be
metaphysically arbitrary. This can be resisted by someone advocating the
brutality of compositional facts (Markosian 1998, 2004*b*). Sider's defense of
P3 proceeds by showing that indeterminacy in composition would result
in *count indeterminacy* in finite worlds—the indeterminacy in the truth-
value of a purely logical sentence stating that there are exactly *n* objects.
Since logical vocabulary, specifically the quantifiers, are not vague (the
assumption questioned by Koslicki 2003; but see Sider 2003, P3 is thereby
vindicated.

Step 2 of the argument from vagueness is modeled after Step 1 but
requires more machinery to state. An *assignment f* is a function from times

---

[16] My explication of these notions and the outline of the argument closely follow Sider 2001:
122–39.

[17] P1 will be rejected by the mereological nihilist, as conceded by Sider 2001: 123. For the remainder
of this discussion, we set nihilism aside.

to non-empty classes of objects existing at those times. A *diachronic fusion* of assignment f (a *D-fusion* of f) is an object x that is a fusion-at-t of $f(t)$ for every t in f's domain. A *minimal D-fusion* of f is a D-fusion of f that exists only at times in f's domain.[18,19] The premises of Step 2 closely parallel those of Step 1:

P1′: If not every assignment has a minimal D-fusion, then there must be a pair of cases connected by a continuous series such that in one, minimal D-fusion occurs, but in the other, minimal D-fusion does not occur.

P2′: In no continuous series is there a sharp cutoff in whether minimal D-fusion occurs.

P3′: In any case of minimal D-fusion, either minimal D-fusion definitely occurs, or minimal D-fusion definitely does not occur.

P1′, P2′, and P3′ imply:

(U) Every assignment has a minimal D-fusion.

For suppose they do not. Then there is a continuous series connecting a case where minimal D-fusion occurs to a case where minimal D-fusion does not occur (P1′). By P3′, there must be a sharp cutoff in this series, which is prohibited by P2′.

But (U) gives the 4Dist what she needs—temporal parts. Consider the assignment $f^* = <t, \{x\}>$, where x is an arbitrary object and t a time at which it exists. By (U), f^* has a minimal D-fusion, z. Using the previously established definition of 'temporal part' and the contrapositive of the temporally qualified version of the strong supplementation axiom of extensional mereology,[20] Sider then shows that z is a temporal part of x at t.[21]

---

[18] Example: Tibbles is a D-fusion of the assignment having two moments in its domain, t_1 (pre-accident) and t_2 (post-accident), and assigning two different but overlapping classes of cells composing it at those times. But Tibbles is not a minimal D-fusion of this assignment because Tibbles exists at times other than t_1 and t_2.

[19] Important: these notions are 3D-friendly; they are based on temporally qualified mereology, which the 3Dist must be happy to embrace. Specifically, 4D is *not* built into the definition of minimal D-fusion.

[20] Sider defines an instantaneous temporal part of y at t as the object x which (i) exists only at t, (ii) is part of y at t, and (iii) overlaps at t everything that is part of y at t (2001: 59). For a more generalized notion, see D4.7^G in §4.1 above. The use of temporally relative notions of *existence-at-t* and *parthood-at-t* makes this definition 3D-friendly. The mereological principle at work states that, if x and y exist at t and x is not part of y at t, then x has a part at t that does not overlap y at t (ibid.: 58).

[21] For details, see Sider 2001: 138–9 and 2003: 136.

4.4. Motivating a Sharp Cutoff

As noted above, other critics reject Step 2 of the argument because they *also* reject Step 1, and for roughly the same reasons. Thus Markosian rejects P2′ because he *also* rejects P2. He notes, rightly, that "a necessary condition for an assignment's having a minimal D-fusion is that the relevant classes of objects all have fusions at the relevant times" (2004*b*: 669). Markosian also indicates that it is not sufficient. How important is this?

I take it to be crucial. A given assignment f could fail to have a minimal D-fusion for two different reasons: (a) the value of f for t (i.e., a given class of objects existing at t) could fail to have a fusion-at-t; or (b) the fusion-at-t_1 of $f(t_1)$ could fail to bear the appropriate relation to the fusion-at-t_2 of $f(t_2)$: identity in the cases of endurance and perdurance, or genidentity in the case of exdurance. This suggests two different strategies for resisting minimal D-fusion universalism. The strategy adopted by Markosian (and, *mutatis mutandis*, by Koslicki) builds on (a) because he already has a reason (i.e., positing brutal compositional facts) to reject unrestricted composition *at a time*. Although he mentions (b), he does not offer any *independent* considerations in its support.

My strategy, on the contrary, is built around (b), and (b) alone. I have nothing to say about (a), which I take to be an advantage. Anyone defending (a) has to deal with a host of difficult issues arising from Peter van Inwagen's "special composition question." I submit, however, that those are quite orthogonal to the issues arising from (b). I insist, therefore, that the reasons for rejecting (DU) may be different from those that might be invoked to resist (SU) and that getting clear on the difference is important for all parties in the debate about persistence.

Recall (§4.2 above) that (SU) is simply the thesis that every class of objects existing at a particular time has a fusion at that time. 4Dists and 3Dists alike should accept this as a fair rendition of unrestricted composition at a time. (DU) is less univocal. In the 4D framework, it is synonymous with the thesis of unrestricted composition *across* time. In the 3D-friendly environment, on the other hand, it amounts to Sider's universalism about minimal D-fusions. Both variants of (DU) can be resisted on the same grounds—the grounds that are different from, and independent of, any considerations having to do with synchronic composition. My chief concern here is to challenge universalism about minimal D-fusions. I turn, accordingly, to what I take to be the problematic premise of Step 2 of the argument:

P2′: In no continuous series is there a sharp cutoff in whether minimal D-fusion occurs.

What would it take to refute it? Short of brutalism (which I unequivocally condemn), one has to produce an example of a continuous series of cases of minimal D-fusion featuring a *motivated* sharp cutoff. To be sure, no attempt at motivating it will find much sympathy with the universalist. But that is irrelevant, for converting the universalist is not on the agenda. What is on the agenda is providing a good rationale for anyone inclined to take issue with a crucial step in an argument aimed at *establishing* (DU) in the first place. It is also worth emphasizing that, since P2′ embodies a *universal* claim about *all* continuous series of putative cases of minimal diachronic fusion, then all that is needed to refute P2′ is to produce a *single* counterexample to it.

I will develop such a counterexample (referred to below as the Example) "backwards," starting with a sharp cutoff and then describing a "continuous series" of cases containing it. The sharp cutoff features a pair of adjacent cases of minimal D-fusion differing in whether minimal D-fusion occurs:

Case 1: $f_1 = \{<t, \{a\}>, t \in T\}$, where T is a's total lifetime

For simplicity, take a to be a mereological atom. (But nothing turns on it; see below.) T is a continuous interval that may be finite or infinite. In Case 1, minimal D-fusion takes place. Indeed, the minimal D-fusion of f_1 is a itself (throughout its entire life career).

Case 2: $f_2 = \{<t, \{a\}>, t \in T'\}$, where $T' = T - \{t^*\}$

In Case 2, minimal D-fusion fails to take place. Motivation: the existence of an object, which would be the minimal D-fusion of f_2, would violate a fundamental law of physics—the law of conservation of matter and energy.[22] Such a fusion would feature an object going out of existence at t^* and popping back into existence *ex nihilo*.[23] As a result, some later phases of such an object would not be connected by a broadly causal relation,

---

[22] In its local form expressed by the continuity equation $\partial\rho/\partial t + \nabla(\rho u) = 0$, where ρ is the local density of matter and u is its local velocity.

[23] If there is the first or last moment of a's existence make an additional stipulation that t^* is distinct from T's endpoint(s).

known in the literature as *immanent causation*,[24] to its earlier phases. No known dynamical laws act across temporal gaps.

Considerations having to do with violating fundamental laws of nature are unlikely to persuade all metaphysicians. The convinced universalist is free to insist that minimal D-fusions such as that of f_2 exist; it's just that they are not the sort of things that physics is interested in (cf. Hudson 2006: §5.2). But such a move will not carry much weight at this point of the dialectic, at which the truth of (DU) has not yet been established. It is the universalist who has to convince the opponent that the opponent's appeal to firmly established results of science and to the notion of immanent causation is *unmotivated*. And on the face of it, it is rather well motivated.

Moreover, one could bring the motivation into sharper relief by noting an important dissimilarity between "unnatural" synchronic and diachronic fusions. It may be an open question whether physics (or any other science) can supply a satisfactory *criterion* for restricting synchronic composition. But in any event, the existence of "unnatural" synchronic fusions cannot be ruled out on any ground of their *inconsistency* with the laws of nature. In contrast, the Example demonstrates that the alleged existence of certain minimal diachronic fusions is strictly *incompatible* with some such laws.

To sum up, Case 1 and Case 2 are two adjacent cases of D-fusion featuring a motivated sharp cutoff in whether minimal D-fusion occurs. (Notice that these cases are as adjacent, and hence the cutoff is as sharp, as they could get: the difference between Cases 1 and 2 is infinitesimal.) Now on to the "continuous series" of cases Σ containing the cutoff in question. Let Σ be the ordered pair <Case 1, Case 2>. Σ satisfies the conditions for being a "continuous series of cases" because (i) Σ is finite, (ii) it is a series of cases of minimal D-fusion, and (iii) any two adjacent cases in Σ are extremely similar in all respects relevant to minimal D-fusion.

---

[24] The general idea of immanent causation is simple: later states of an object generally depend on its earlier states, roughly in a way in which the state of one object causally depends on prior states of another object that causally affects the first. This language is supposed to be neutral with respect to the controversy about persistence. The 4Dist is free to regard "later and earlier states of an object" as the states of distinct object stages, whereas the 3Dist is free to regard them as the states of the selfsame object. Simple though it is, the general idea of immanent causation is difficult to make precise. But most of the difficulties come from 'causation,' not from 'immanent.' One such difficulty is that causal connectedness may vary in degree depending on the temporal distance between the corresponding object states and perhaps vanish altogether for states that are very far apart. For a comprehensive analysis of immanent causation in terms of nomic subsumption of events, see Zimmerman 1997.

The existence of Σ shows that P2$'$ is false and, hence, Step 2 of the argument from vagueness is unsound.

4.5. Some Objections and Replies

Objection 1: Using the unit class $\{a\}$ of a single object trivializes the issue of whether that class has a fusion at a time, for a unit class *always* has a fusion (cf. Sider 2001: 123, n. 46). As a result, we do not have a continuous series connecting a case of composition with a case of *non*-composition.

Reply 1: But we do have a continuous series connecting a case in which minimal D-fusion occurs with a case in which it does not. Our choice of f_1 and f_2 does *not* trivialize the issue of whether such assignments have minimal *diachronic* fusions. The Example shows that the nature of such fusions is different from the nature of composition at a time. I say that the former, but not the latter, includes broadly causal relations which figure prominently in the fundamental physical laws.

Reply 2: Nothing of substance turns on the choice of a single object. The Example could be modified to include a class of n objects existing throughout T and such that, at any time during T, they compose (at that time) a certain object a. For the same reason, nothing substantial turns on a's being a mereological atom.

Objection 2: The Example is weird in that it involves a "continuous series of cases" with only two members. This is not what one expects in a discussion of vagueness. What one normally expects is a "sufficiently long" series of cases with the relevant sort of change occurring somewhere in the "middle."

Reply 1: This expectation comes from *Sorites* cases, which may serve as a good model for synchronic composition—because the latter may be alleged to turn on small differences in the multigrade relations among many objects *at a time* (e.g. proximity, contact, chemical bonding, etc.)—but not necessarily for diachronic relations, because, as noted above, those relations have a different nature having to do with immanent causation and restrictions imposed by the laws of nature.

A Sorites series of cases of synchronic composition would begin with a clear case of composition (a compact and well-formed collection of atoms, such as an animal body) and end with an alleged case of non-composition (the collection in question scattered throughout the universe). The universalist would then argue that any sharp cutoff "somewhere in the middle" would be metaphysically arbitrary and, hence, all synchronic fusions should be allowed, no matter how scattered and gerrymandered their material is. She could add that the existence of such fusions would not *violate* any physical laws.

Not so in the case of diachronic fusions, where a motivated sharp cutoff can be associated with the *very first* tiny deviation from a clear case of D-fusion, such as one that occurs in the transition from Case 1 to Case 2. Deviations of this sort *immediately* violate the laws of nature. Thus it should not be surprising that the cutoff occurs at the "very first step" and, hence, no long series of cases is needed to illustrate it.

Reply 2: Nothing in the argument itself *requires* that a continuous series be "sufficiently long": P2′ universally quantifies over *all* continuous series.[25]

Reply 3: One is free to turn Σ into a "sufficiently long series" by adding more cases of the same sort (i.e., by excluding from T more and more instants). The cutoff would still occur at the very first step. And this should still be recognized as a natural feature of a physically motivated restriction on diachronic fusions. The upshot is that Σ is a legitimate series and, thus, a counterexample to premise P2′ of the argument.

Objection 3: Examples such as the Example could be excluded by imposing topological constraints on the domain of f, say, by restricting such domains to continuous intervals or sums of continuous intervals of time (cf. Sider 2001: 136). Restrictions of this sort would affect the argument, but the argument thus affected "would still establish a restricted version of [4Dism] according to which there exist continuous temporal segments of arbitrarily small duration. For most [4Dists] that would be [4Dism] enough [ibid.]."

Reply 1: (a) I do not see immediately how the argument thus modified would still establish a version of 4Dism; and Sider does not elaborate. (b) Even if it would, I believe the Example could be modified in a similar way.

Reply 2: But we need not quarrel, for invoking topological restrictions on intervals of time is, in the present context, a red herring. The Example

---

[25] Incidentally, the assignments f_1 and f_2, which are responsible for the allegedly "weird" character of the Example because they incorporate a single object and a single moment of time eliminated from T, are not unlike the assignment f^* in the (indispensable) Step 3 of the argument from vagueness, which also incorporates a single object and a single moment of time.

is based on violating a *physical* law, and not on any topological property of time. To set these two issues apart, let us modify Case 2 as follows:

Case 2′: $f_2' = \{<t, \{a\}>, t \in T'\} \cup <\{t^*, \{b\}>,$

where (i) $T' = T-\{t^*\}$, (ii) $b \neq a$, (iii) b has the same intrinsic physical properties as a, (iv) b is located as close as physically possible to a.

In Case 2′, minimal D-fusion fails to take place, for the existence of an object, which would be the minimal D-fusion of f_2', would violate the laws of motion. Such a fusion ('*aba*,' to give it a name) would feature an object "jerking" for an instant from its lawful trajectory. Relatedly, some later phases of *aba* would not be connected by immanent causation to some of its earlier phases. Cases 1 and 2′ are truly adjacent, even if not infinitesimally close. Consequently, just like Σ, the series $\Sigma' = <$Case 1, Case 2′$>$ exemplifies a motivated sharp cutoff in whether D-fusion occurs. But the domain of f_2' is now a continuous interval of time.

4.6. Implications

The argument from vagueness to 4Dism fails. Let us take a closer look at the general dialectical situation surrounding it. The argument bases the doctrine of temporal parts on a prior endorsement of unrestricted composition, along the spatial as well as temporal dimensions. Not only is there a momentary object consisting of my nose tip and the Eiffel Tower, but there is also a temporally extended object fusing earlier phases of Napoleon's favorite dog with later phases of the Golden Gate Bridge. To get any mileage from the argument, the 4Dist must accept such monsters into her ontology. But isn't this price too high? As a matter of sociology, many 4Dists accept the bargain. The question is whether they have to.

If the above considerations are sound, they do not. More precisely, the 4Dist is free to treat synchronic and diachronic composition differently. The latter may be causally grounded in a way the former is not. Even in the absence of any principled criterion of restricted composition at a time, the 4Dist can draw a distinction between series of object stages cemented by a broadly causal relation and those that are not. Thus defending some version of restricted synchronic composition is not a prerequisite for resisting diachronic universalism. Accordingly, there is no pressure for the 4Dist to be a universalist across the board. She is free to *banish*, not merely disregard, loose collections of unrelated temporal parts (such as certain parts of *aba*) even if she is not yet prepared to do away with loose collections of spatial parts. While some may view such partial rejection of universalism as arbitrary, I submit that it simply follows the joints of nature. Broadly causal relations linking the state of a particle (or of its temporal part) at one time to its state (or to the state of its other temporal part) at a later time are physically determinate, even if they are not observationally perspicuous, and even if they are normally enmeshed with their more problematic synchronic cousins—for example, when we wonder whether a fusion of a certain class of particles at t_1 is the same object as the fusion of a different such class at t_2. Particle-by-particle worms can still be disentangled from the mess, and the question of which *spatially* composite objects are *also* diachronically related is a *separate* issue—an issue in *synchronic* composition, which, in this case, boils down to the question of which composite objects comprise which collections of particles at *particular* moments of time.

Despite the failure of the argument from vagueness, the overall case for 4Dism, I contend, remains strong. For example, 4Dists' ability to provide a unified solution to the paradoxes of coincidence may still count as a persuasive reason in its favor, and the cumulative weight of other arguments for 4Dism is significant.[26] Moreover, the considerations of this chapter have been confined to the classical framework. We have not yet fully tapped the resources of contemporary spacetime theories in physics, which appear to be highly relevant to the issue of persistence. It is time to do so. Incidentally, the notion of immanent causation, which has come to light in the discussion of the argument from vagueness, will have a role to play in the sequel.

---

[26] See Sider 2001: ch. 4 for an extensive array of some such arguments.

5

Persisting Objects in Minkowski Spacetime

Up until the last decade the debate about persistence was conducted, for the most part, in complete abstraction from physics.[1] The situation has now changed, with a growing number of works exploring various implications of relativity theory for the ontology of persistence.[2] There are two allegedly separate tasks here: (a) to state the rival views in the relativistic context; (b) to investigate whether such statements privilege a particular view over its rivals. There is considerable disagreement as regards both (a) and (b). Indeed, some authors have contended that certain views of persistence cannot even be stated in the relativistic framework. While certain considerations to this effect can be shown to arise from confusion about what relativistic persistence properly amounts to, others are highly sophisticated. I shall examine such proposals later in the chapter. In the next section I sketch one feasible approach to defining different modes of persistence in Minkowski spacetime, noting, along the way, certain assumptions underlying this approach, which have been perceived by others as controversial. I thus set myself in opposition to those who object to the very possibility of relativistic formulations of the major views of persistence. My strategy builds on the classical ideas developed in §4.1 and the geometrical approach to spacetime theories outlined in Chapter 3.[3]

---

[1] Notable exceptions include Quine 1960: 171–2, 1987 and Smart 1972. I discuss Quine's and Smart's early considerations in §5.3 below.

[2] See Rea 1998; Balashov 1999, 2000a, b, c, 2002, 2003a, b, 2005c, 2008; Sider 2001: §4.4; Gilmore 2002, 2006, 2007, 2008; Hales and Johnson 2003; Miller 2004; Hudson 2006: ch. 5; Gibson and Pooley 2006; Eagle 2009a.

[3] For related approaches see Rea 1998; Sider 2001: §4.4; and Sattig 2006: §5.4. For a critique of such approaches, see Gibson and Pooley 2006. I address the critique in §5.2.

5.1. Enduring, Perduring, and Exduring Objects in Minkowski Spacetime

In classical spacetime, locations (and q-locations) of persisting objects, their parts and counterparts, as well as temporary properties were indexed by moments of absolute time—or, more precisely, by t-slices of the objects' paths (see §4.1). A natural adaptation of such indexing to the special relativistic framework requires further relativization to inertial frames of reference. This suggests a straightforward way of extending the definitions of §4.1 to Minkowski spacetime (ST^M) by replacing the classical 't' with a two-parameter index 't^F' tracking moments of time in a particular coordinate system adapted to a given inertial frame F.

A bit more formally, let us identify *absolute chronological precedence* in ST^M with the frame-invariant relation in which two points stand just in case they are either timelike separated or lightlike separated while being distinct: $p_1 < p_2 \leftrightarrow I(p_1, p_2) \geq 0 \wedge p_1 \neq p_2$, where $I(p_1, p_2) \equiv c^2(t_2 - t_1)^2 - (x_2 - x_1)^2$ is the relativistic interval. Accordingly, any spacelike hypersurface[4] counts as an achronal region of ST^M:

(D5.1^M) Region R of ST^M is *achronal* $=_{df} \forall p_1, p_2 ((p_1, p_2 \in R \wedge p_1 \neq p_2) \rightarrow I(p_1, p_2) < 0)$.

But only a subset of them—those that are *flat*—represent legitimate perspectives: moments of time in inertial reference frames, $\{t^F\}$:

(D5.2^M) R is a *moment of time* in ST^M $=_{df}$ R is a spacelike hyperplane in ST^M.

It is therefore appropriate to index q-locations of persisting objects and of their parts in ST^M to t^F.[5] And it is convenient to treat 't^F' as a two-parameter index, assuming that the choice of a particular coordinate system adapted to a given inertial reference frame can be fixed.[6]

---

[4] A hypersurface is spacelike just in case any two points on it are spacelike separated.

[5] The appropriateness of restricting "legitimate perspectives" and q-locations of persisting objects and their parts to moments of time in inertial reference frames in ST^M has been criticized by Gibson and Pooley 2006: 159–65. I discuss and respond to their criticism in the next section.

[6] A reminder (see §4.1): The definitions of this section are *adaptations* of the generic definitions of §2.4 to Minkowski spacetime, which involves systematic replacement of all the relevant generic notions

Two related facts about frame-relative moments of time in ST^M are worth noting:

(i) Any two distinct moments of time $t^F{}_1$ and $t^F{}_2$, $t^F{}_1 \neq t^F{}_2$, in a single frame F are parallel and, therefore, do not overlap. In this respect, moments of time in a given frame are similar to absolute moments of time in ST^G.

(ii) Any two moments of time in distinct frames, $t^{F_1}{}_1$ and $t^{F_2}{}_2$, $F_1 \neq F_2$, overlap. In this respect, moments of time in distinct frames in ST^M are very different from absolute moments of time in ST^G.

Q-location and *path* in ST^M can then be defined.

(D5.3^M)   *o* is (exactly) *q-located* at region R of ST^M $=_{df}$ one of *o*'s (non-modal) counterparts is (exactly) located at R.[7]

(D5.4^M)   Spacetime region *o* is the *path* of object *o* in ST^M $=_{df}$ *o* is the union of the spacetime region or regions at which *o* is q-located.

On the generic definition of persistence (D2.5) from §2.4.3, *o persists* just in case *o*'s path is non-achronal. In Minkowski spacetime, this is equivalent to the requirement that *o*'s path intersect at least two distinct moments of time in a single frame or, alternatively, that *o*'s path contain two non-spacelike separated points.

(D5.5^M)   *o* persists in ST^M $=_{df} \exists p_1, p_2 \in o$ $(\exists F\ t^F{}_1 \neq t^F{}_2)$ $=_{df} \exists p_1, p_2 \in o$ $(p_1 \neq p_2 \wedge I(p_1, p_2) \geq 0)$.

As before, $(x^F{}_1, t^F{}_1)$ and $(x^F{}_2, t^F{}_2)$ are the coordinates of p_1 and p_2 in a Cartesian coordinate system adapted to the inertial frame of reference F.

As we did in the classical case (§4.1), we can adapt the generic definitions of achronal (D2.6) and diachronic (D2.7) parthood of §2.4.4 to Minkowski spacetime by restricting our consideration to achronal regions of interest.

with their Minkowskian counterparts (e.g., *achronal* gives way to *achronal-in-ST^M*, etc.). To simplify the notation, such replacement is presupposed without making it in each case explicit.

[7] Alternatively (cf. n. 1 in §4.1):

(D5.3$^{M'}$)   *o* is (exactly) *q-located* at region R of ST^M $=_{df}$ *o* is (exactly) located at R or one of *o*'s (non-modal) counterparts is (exactly) located at R.

In particular, an achronal slice $R_\perp$ of R in ST^M is the intersection of R with a moment of time in some inertial frame:

(D5.8^M) $R_\perp$ is an *achronal slice* of R in $ST^M =_{df} R_\perp$ is a non–empty intersection of a moment of time (i.e., a time hyperplane) with R.

We shall refer to the achronal slice of R at t^F in ST^M as 't^F-slice of R' or '$R_{\perp t^F}$'.

As before, we start with a primitive three–place relation 'p is a part of o at achronal region R.' Achronal regions of interest are now t^F-slices of the objects' paths. Where p, o and a t^F-slice $o_{\perp t^F}$ of o's path o stand in such a relation, we shall say that p is an *achronal part* (s^F-*part*) of o at $o_{\perp t^F}$:

(D5.6^M) $p_\perp$ is an *achronal part* (s^F-*part*) of o at $o_{\perp t^F}$ in $ST^M =_{df} p_\perp$ is a part of o at $o_{\perp t^F}$.

And we adapt the generic notion of diachronic parthood as follows:

(D5.7^M) $p_\parallel$ is a *diachronic part* (t^F-*part*) of o at $o_{\perp t^F}$ in $ST^M =_{df}$ (i) $p_\parallel$ is located at $o_{\perp t^F}$ but only at $o_{\perp t^F}$, (ii) $p_\parallel$ is a part of o at $o_{\perp t^F}$, and (iii) $p_\parallel$ overlaps at $o_{\perp t^F}$ everything that is a part of o at $o_{\perp t^F}$.

For convenience, we shall allow such expressions as 'achronal part of o at t^F,' 'diachronic part of o at t^F' and 'o's t^F-part' to go proxy for their more complex equivalents, such as 'achronal part of o at t^F-slice $o_{\perp t^F}$ of o's path o' and so forth, where context makes it clear that 't^F' refers not to an entire time hyperplane but to a rather small subregion of it—$o_{\perp t^F}$. Moreover, we shall sometimes allow ourselves the liberty to speak of "spatial" and "temporal" parts of persisting objects in Minkowski spacetime when it is clear what reference frame is under consideration.

As before, we introduce the notion of *object-eligible achronal slice* of the object's path[8] appropriate to Minkowski spacetime:

(D5.11^M) $o_{\perp t^F}$ is an *o-eligible t^F-slice of o's path* $o =_{df}$ either o itself or o's t^F-part at $o_{\perp t^F}$ is q-located at $o_{\perp t^F}$.

The definitions of the three basic modes of persistence in Minkowski spacetime are then as follows:[9]

---

[8] See §2.4.5: cf. D2.11^G from §4.1.
[9] Cf. their classical predecessors D4.12^G–D4.14^G from §4.1.

(D5.12^{M}) *o endures* in ST^{M} =$_{\mathrm{df}}$ (i) *o* persists, (ii) *o* is located at every *o*-eligible t^{F}-slice of its path, (iii) *o* is q-located only at *o*-eligible t^{F}-slices of its path.

(D5.13^{M}) *o perdures* in ST^{M} =$_{\mathrm{df}}$ (i) *o* persists, (ii) *o* is q-located only at its path, (iii) the object located at any *o*-eligible t^{F}-slice of *o*'s path is a proper t^{F}-part of *o* at that slice.

(D5.14^{M}) *o exdures* in ST^{M} =$_{\mathrm{df}}$ (i) *o* persists, (ii) *o* is located at exactly one region, which is an *o*-eligible t^{F}-slice of its path, (iii) *o* is q-located at every *o*-eligible t^{F}-slice of its path.

The above definitions need to be supplemented with an account of the relativization of temporary properties of persisting objects to their locations (in the case of endurance), q-locations (in the case of exdurance), or the locations of their t^{F}-parts (in the case of perdurance). Such (q-) locations are, of course, t^{F}-slices of the objects' path, which can be usefully labeled with the same two parameter-index that figures in the above definitions.

A bit more formally, the analysandum of the relativization schemes characteristic of endurance, perdurance, and exdurance in ST^{M} is an expression of the form '*o* has Φ at t^{F}' (where, as before, 't^{F}' is shorthand for what, in a more formal notation, would be a complex index '$o_{\perp t^{\mathrm{F}}}$').[10]

(D4.15^{M}) Enduring object *o* has Φ at t^{F} (i.e., at $o_{\perp t^{\mathrm{F}}}$) in Minkowski spacetime =$_{\mathrm{df}}$ *o* bears Φ-at to t^{F}.

(D4.16^{M}) Enduring object *o* has Φ at t^{F} (i.e., at $o_{\perp t^{\mathrm{F}}}$) in Minkowski spacetime =$_{\mathrm{df}}$ *o* has Φ-at-t^{F}.

(D4.17^{M}) Enduring object *o* has Φ at t^{F} (i.e., at $o_{\perp t^{\mathrm{F}}}$) in Minkowski spacetime =$_{\mathrm{df}}$ *o* has $_{t^{\mathrm{F}}}$ Φ.

(D4.18^{M}) Perduring object *o* has Φ at t^{F} (i.e., at $o_{\perp t^{\mathrm{F}}}$) in Minkowski spacetime =$_{\mathrm{df}}$ *o*'s t^{F}-part has Φ.

(D4.19^{M}) Exduring object *o* has Φ at t^{F} (i.e., at $o_{\perp t^{\mathrm{F}}}$) in Minkowski spacetime =$_{\mathrm{df}}$ *o*'s t^{F}-counterpart has Φ.

Thus, while in the Galilean framework objects have properties at absolute moments of time (more precisely, at absolute time slices of the objects'

---

[10] D4.15^{M} describes, on behalf of the endurantist, (Minkowskian) relationalism (not to be confused with spacetime relationism), D4.16^{M} (Minkowskian) indexicalism, and D4.17^{M} (Minkowskian) adverbialism. Cf. their Galilean predecessors D4.15^{G}–D4.17^{G} from §4.1.

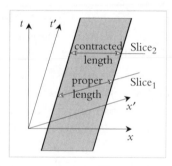

Figure 5.1. 2-meter stick in Minkowski spacetime.

paths; see the end of §4.1), in the Minkowskian framework the possession of temporary properties is relativized, in effect, to times-in-frames. This brings some novel features. Consider, for example, a 2-meter stick, whose path is a shaded region in Figure 5.1 (with two dimensions of space suppressed). Even if the stick does not change its *proper length* (i.e., the length it has in its rest frame), it exemplifies different length at time slices of its path drawn in different reference frames, such as Slice$_1$ and Slice$_2$. The endurantist will say that the stick is located at both slices and bears the *2 meters-at* relation to Slice$_1$ and *1 meter-at* relation to Slice$_2$.[11] The perdurantist will say that the stick is located at its path and has two distinct t^F-parts at Slice$_1$ and Slice$_2$ of different corresponding length. The exdurantist will say that the stick is q-located at Slice$_1$ and Slice$_2$ and has simple lengths *2 meters* and *1 meter* there, thanks to its counterparts. The whole situation is familiar and illustrates the relativistic phenomenon of length contraction (cf. §3.6).

5.2. Flat and Curved Achronal Regions in Minkowski Spacetime

In the generic spacetime framework introduced in Chapter 2, locations and quasi-locations of persisting objects were indexed to arbitrary achronal

---

[11] Alternatively, the endurantist can say (i) that the stick atemporally or tenselessly exemplifies two *timeslice-indexed* lengths, *2 meters-at-Slice$_1$* and *1 meters-at-Slice$_2$*; or (ii) that the stick exemplifies two simple length properties, *2 meters* and *1 meter*, in two different *timeslice-modified ways*. Cf. the corresponding classical analyses of §4.1.

regions. The adaptation of the general definitions of the different modes of persistence (and a host of attendant notions) to Minkowski spacetime in the previous section was based on the assumption[12] that persisting objects and their parts are q-located (and, consequently, have properties) at *flat* achronal regions representing, in special relativity, moments of time in inertial reference frames. Let us explicitly refer to this assumption as FLAT:

> (FLAT) In the context of discussing persistence in Minkowski spacetime it is appropriate to restrict the locations and q-locations of persisting objects and their parts to flat achronal regions representing (subregions of) moments of time in inertial reference frames.

Initially one might be inclined to reject FLAT on rather general metaphysical grounds. Consider a *non*-flat achronal slice $o_\perp$ of object o's path in Minkowski spacetime. How could o (if o endures or exdures), or one of o's diachronic parts (if o perdures), *fail* to be q-located at $o_\perp$? In other words, how could $o_\perp$ fail to "contain" o (or one of o's diachronic parts)? After all, $o_\perp$ is an achronal slice of o's path and is matter-filled; therefore it must contain *something*! And what could this "something" be except o or one of o's diachronic parts?

This general line of thought should be resisted (cf. Gilmore 2006: 210–11), because in turns on conflating the notion of an achronal region's being a q-location of o (or one of its diachronic parts) with the notion of an achronal region's being "filled with achronal material components of o." A region may satisfy the latter property without satisfying the former. Imagine Unicolor, a persisting object one of whose essential properties is to be *uniformly colored*. Suppose further that Unicolor uniformly changes its color with time in a certain inertial reference frame F. Consider an achronal slice of Unicolor's path, flat or not, that crisscrosses hyperplanes of simultaneity in F. Whatever (if anything) is q-located at such a slice is not uniformly colored and, hence, must be distinct from Unicolor, even though it is filled with the (differently colored) achronal material components of Unicolor.[13]

---

[12] Shared by a number of other writers; see, in particular, Sider 2001: 59, 84–6; Rea 1998; Sattig 2006: §§1.6 and 5.4. Gilmore, who held this assumption in his earlier work (Gilmore 2004), appears to have abandoned it later; see, in particular, Gilmore 2006. Unlike Gibson and Pooley 2006, however, he does not offer any specific criticisms of the assumption.

[13] But compare a critical discussion of a similar example due to Smart in §5.3 below.

We shall return to the issue of crisscrossing slices in §§5.5 and 5.6 below. For the moment, let us simply note that, in view of the Uni-color example, general metaphysical considerations are not sufficient to reject FLAT. But notice that the property of being *uniformly colored* used in the example is itself grounded in a prior concept of spatial or achronal *uniformity*, which, in turn, presupposes that flat achronal regions of Minkowski ST are somehow physically privileged in the context of SR. In a recent work Ian Gibson and Oliver Pooley (2006: 160–5) have argued against this view, thereby presenting a more pointed objection to FLAT. Their objection also raises important methodological questions about the relationship between physics and metaphysics. Below I consider and respond to Gibson and Pooley's objection and, in the course of doing so, address the methodological concerns brought to light in their critique of FLAT.[14]

In Gibson and Pooley's view, the tendency to "frame-relativize" in the manner of FLAT and other similar assumptions, which is adopted unreflectively by several authors discussing persistence in the context of Minkowski spacetime (see note 12 above), represents a relic of the classical worldview and stands in the way of taking relativity seriously. While inertial frames of reference (i.e., spacetime coordinate systems adapted to them) are geometrically privileged and, therefore, especially convenient for describing spatiotemporal relations in Minkowski spacetime, this does not give them any distinguished metaphysical status. Accordingly (and contrary to FLAT), no such status should be granted to flat achronal regions in Minkowski spacetime. Thus Gibson and Pooley:

From the physicist's perspective, the content of spacetime is as it is. One can choose to describe this content from the perspective of a particular inertial frame of reference (i.e., to describe it relative to some standard of rest and some standard of distant simultaneity that are optimally adapted to the geometry of spacetime but are otherwise arbitrary). But one can equally choose to describe the content of spacetime with respect to some frame that is not so optimally adapted to the geometric structure of spacetime, or indeed, choose to describe it in some entirely frame-independent manner.   (Gibson and Pooley 2006: 162)

<hr/>

[14] Gibson and Pooley' objection to FLAT also prepares the ground for a subsequent attack on one of my arguments favoring perdurance over endurance and exdurance in the context of SR; see Balashov 2000a, b, 2002, and 2005. I present a modified version of this argument in Chapter 7 below, where I also respond to my critics: Gilmore 2002; Miller 2004; and Gibson and Pooley 2006.

. . .

More significantly, one surely wants a definition [of a notion relevant to char-
acterizing a mode of persistence in spacetime—Y. B.] applicable in the context
of our best theory of space and time, general relativity. While this theory allows
spacetimes containing flat spacelike regions, generic matter-filled worldtubes will
have *no* flat maximal spacelike subregions. The obvious emendation, therefore,
is simply to drop clause (iv) [i.e. FLAT or some analogous assumption—Y. B.].
(Ibid., 163)

These remarks contain two distinct points, and both raise important questions. The first point—that inertial reference frames and flat regions in
Minkowski spacetime are privileged only geometrically and not physically
and, therefore, do not warrant ascribing to them any metaphysical significance in the context of questions about persistence—appears to derive its
force from a crucial lesson of the contemporary methodology of spacetime
theories: that the choice of a local coordinate system is completely arbitrary
and has no bearing whatsoever on the content of a particular spacetime
theory. Any such theory—Newtonian mechanics, classical electrodynam-
ics or special relativity—can be formulated in any coordinate system.
Moreover, such a formulation can always be made covariant with respect
to arbitrary local coordinate transformations, at the cost of making it less
elegant.

For example, Newtonian mechanics of free particles in Galilean spacetime
can be stated in terms of a set of geometrical objects on the manifold[15]
satisfying certain field equations and the equation of motion that are
expressed in arbitrary local coordinate systems.[16] It turns out that there is
a special sub-class of *inertial* coordinate frames, in which the equation of
motion takes the familiar form of Newton's first law. Although this fact
obviously has enormous practical significance, it allows us to use a simple
expression of Newton's first law in a great variety of practical applications,
the fact that such frames exist has no physical importance.

Indeed, suppose a certain particle performs a non-inertial motion. One
could then associate with it a series of instantaneous rigid Euclidean systems
that, intuitively, "track" its motion and allow one to treat the particle as

[15] Specifically, an affine connection, a covariant vector field, and a two-rank symmetric tensor. My
informal outline of this example follows Friedman 1983: 87–94.
[16] Alternatively, as Gibson and Pooley note, such equations can also be given a coordinate-free
formulation.

being, effectively, at rest, at the cost of introducing fictitious forces (namely, the "inertial force" and the "Coriolis force"). The point to note here is that, in the end, the presence of straight non-achronal "position lines," which allow one to identify spatial positions at different times in perspectives associated with inertial coordinate systems, has no physical consequence: curved lines (those that track the motion of non-inertial particles) would do just as well. Based on this point, one could argue that position in space, as defined in a *given* inertial frame, is a rather thin notion that hardly bears the weight attributed to it in many metaphysical discussions—even in the context of classical physics.

And things get worse. Even in *that* context, one can choose to "geometrize away" gravitational forces by incorporating the gravitational potential into the affine connection, at the cost of making the classical spacetime non-flat (i.e., by making it curved).[17]

This example shows that, *even in the classical context*, the presence of a well-defined family of straight diachronic position lines and the usual assumption that spacetime as a whole is flat have no physical significance. Does this mean that one should ban familiar notions, such as *same place over time in a given inertial frame*, from philosophical discussions tailored to the classical context, simply because inertial frames and straight achronal lines enjoy no special status at the fundamental level of physical description?

Hardly so. Banning such notions would deprive one of many useful resources in a situation where such resources are *available*. Note that the issue does not concern the retention of the notion of *sameness of place over time*, period (even the classically minded metaphysician can be convinced that the latter notion *is* meaningless, see §3.2), but only the significance of the notion of sameness of place over time *in an inertial frame*. This notion provides resources for imposing a global coordinatization on spacetime and assigning various conceptual roles to such a coordinatization. It would appear that the metaphysician should feel free to make use of the salient concept of sameness of place across time (against the backdrop of a particular inertial frame)—as long as such a concept is definable—even if physics, in the end, denies distinction to inertial frames.

Two facts seem to be relevant here: (i) that global inertial coordinate systems are *available* (despite the lack of physical importance) and (ii) that their

---

[17] We shall not pursue this further. See Friedman 1983: 95–104 for details.

availability allows one to *minimize revision* of the existing ontological vocabulary. The above brief excursus into Newtonian mechanics should serve to support (i). (ii), on the other hand, raises more general considerations having to do with philosophical methodology.

It is a well-known fact that most contemporary discussions in fundamental ontology[18] continue to be rooted in the "manifest image of the world" and ignore important scientific developments, which have rendered many common-sense notions untenable and obsolete. Attempts to bring physical considerations to bear on issues in fundamental ontology, such as those discussed in this book, are still very rare. This persistent self-isolation of contemporary metaphysics from science may prompt two different reactions from philosophers who are wary of "armchair philosophical speculation." One may be tempted to reject such speculation, root and branch, and adopt the following attitude: let physics tell us what the world is like and then let the "metaphysical chips" fall where they may. It is unclear whether any part of the contemporary metaphysical agenda would survive such a treatment. But it is equally unclear whether any consistent worldview could emerge from it. Science is an open-ended enterprise which is becoming increasingly fragmented. The same is true of any particular scientific discipline, such as physics. The question of what parts of contemporary fundamental physics could contribute safe and reliable components to the foundations of an overall worldview is a highly complex question, which may not have a good answer.

This suggests a different attitude. One may admit that the prolonged mutual alienation of metaphysics and physics is unfortunate but insist that both have some value in their *current* state, and could therefore benefit from gradual rapprochement. It should be clear that the present study follows the second course. It should also be clear, by now, that this course brings with it various limitations and requires certain methodological decisions to be made at some junctures.[19]

One sort of limitation has to do with the restricted domain of physical possibility, which renders certain metaphysical scenarios (including some philosophy favorites) irrelevant. If one wants to see what bearing a particular physical theory may have on a metaphysical debate, one should be

---

[18] That is, discussions of such issues as time, persistence, material composition, the nature of fundamental properties and laws, etc.

[19] Some of these were set out in Chapter 1.

firmly grounded in the set of possibilities allowed by the theory under consideration.

Another limitation has to do with the choice of the physical theory (or theories) under consideration. Given the open-ended nature of physics any physical theory is likely to be false. But one hopes that some theories are good approximations to the truth, and to the extent that they are, adapting existing metaphysical views to them is valuable.[20] The scope of the present study is restricted, for the most part, to special relativity. This represents a particular choice and brings with it quite obvious restrictions.

Even more important, when engaged in extending an existing meta-physical debate to a new physical framework one confronts non-trivial judgment calls at many turns, when it becomes clear that some familiar notions must be abandoned, others modified, while others can be kept more or less intact. Usually there is more than one way to "save the appearances," but the decision as to what "intuitions" must be retained at the expense of others is difficult because one is now swimming in uncharted waters. In the end, it is the entire resulting systems and their performance across a variety of theoretical tasks that must be compared. I submit that the only reasonable regulative maxim to be imposed on physically informed metaphysical theorizing should be stated in terms of *Minimizing Overall Ontological Revision* (MOOR). Vague as it is, its role could be favorably compared to Quine's famous criteria of "conservatism," "the quest for simplicity," and "considerations of equilibrium" affecting the "web of belief as a whole":

> (MOOR)   In adapting a metaphysical doctrine to a physical theory one should seek to minimize the degree of the overall ontological revision.

As we depart from the "comfort zone" of the classical worldview, the degree and extent of the "overall ontological revision" become progressively up for grabs, which makes MOOR increasingly wholesale and non-specific. But as indicated above, any alternative to MOOR would amount to rejecting the entire agenda of contemporary metaphysics. I should emphasize that the latter is not what Gibson and Pooley undertake to do in the above-quoted work (Gibson and Pooley 2006). Having noted

---

[20] But see Monton MS for objections to such a strategy.

that they have "a lot of sympathy" for the view that "the project of reconstructing relativistic versions of familiar non-relativistic doctrines [may be] horribly misguided,"[21] they "nonetheless think that it is worthwhile to engage with attempts to square the familiar doctrines with relativity" (ibid.: 157–8). Such attempts, I recommend, must be guided by something like MOOR.

Returning (finally) to FLAT, I contend that it conforms to the spirit of MOOR quite well. Indeed FLAT employs structures (namely, global flat hypersurfaces) that are (i) available in Minkowski spacetime, (ii) widely used in physics, and (iii) are indispensable to extending the important notions of moment of time and momentary location of an object or its part (in a given reference frame) to the special relativistic framework. In this respect, FLAT is on a par with the license to attribute importance to a family of straight position lines in classical spacetime despite the fact that, at bottom, straight diachronic lines do not enjoy (*even* in Galilean spacetime) any physical privilege over curved diachronic lines. The important facts are that (i) straight lines are *definable* in that context and that (ii) without their presence, the notion of "place over time in a given frame" would get completely out of touch with any familiar notions. For similar reasons, global hyperplanes can be assigned important roles in Minkowski spacetime. First, they are easily definable as such. Second, if they lose their privilege over arbitrary achronal hypersurfaces vis-à-vis issues of persistence, the notion of *momentary* location of a persisting object—and, with it, the host of other notions tied up to momentary location, such as momentary shape, momentary achronal composition, and the like—would lose much of their ground and would be hard to anchor in any familiar concepts. They would become too remote to perform any meaningful function in a metaphysical debate.

I conclude that FLAT is justified in the context of Minkowski spacetime. But I fully agree with Gibson and Pooley that it is not appropriate for *general* relativistic spacetime, where matter-filled flat achronal regions are *not available*. Since that context has no place for global "moments of time" and "momentary locations," the connection with the familiar set of notions is severed anyway and there is no pressure to align other concepts with

---

[21] "Should we not start with the relativistic world picture and ask, in that setting and without reference to non-relativistic notions, how things persist?" (Gibson and Pooley 2006: 157–8)

them. In general relativistic spacetime it may be natural to regard any achronal slice of an object's path as a good candidate for the object's (or its part's) location—if one thinks that the notion of location continues to make any sense there.

My consideration is generally limited to Minkowski spacetime of special relativity, which, for the purpose of discussion, is taken to be a good approximation of the spacetime of our real world. Even so, the issue of the status of curved hypersurfaces in Minkowski spacetime goes beyond the above brief remarks. Some facts about such hypersurfaces are non-trivial and interesting in their own right. We shall revisit this topic in Chapter 7 (§7.9 and Appendix 7.1).

5.3. Early Reflections on Persisting Objects in Minkowski Spacetime: Quine and Smart

Even the brief description of §5.2 makes it clear that the Minkowski framework introduces new and counterintuitive features into the manner in which persisting objects can be said to exist in spacetime. This has not escaped the attention of earlier commentators. Indeed, some of them exploited such peculiar features to argue that special relativistic spacetime *requires* a certain mode of persistence—namely, perdurance—and rules out, or at least strongly disfavors, endurance. Although these early works did not use a consistent terminology similar to one with which we are now familiar, the claims actually made do not leave any doubt as to what account, or picture, of persistence their authors advanced on each particular occasion. This section reviews some of these early relativistic considerations.

Before we turn to them it should be noted that although many discussions of the four-dimensional ontology of material objects were undoubtedly stimulated by the advent of SR and GR,[22] the ontology of 4Dism has a much older history.[23] Putting aside the problem of change, the topic of vagueness, the riddles of coincident entities, and the impact of contemporary physical

---

[22] See, in particular, Whitehead 1920 and Russell 1927.

[23] Which is still awaiting its scholar. Hume is probably the most famous modern precursor of 4Dism. Sider mentions McTaggart, Hermann Lotze, Jonathan Edwards, Heraclitus and some Buddhist doctrines among others (2001: 3, n. 3). This list is surely far from complete.

theories—the issues which figure in present-day debates—the notion of persistence via temporal parts can be motivated by considerations as diverse as skepticism about strict identity or identification across time,[24] the alleged conceptual supremacy of tenseless predication in logic,[25] and various analogies between space and time.[26] Some of these issues are very old.

One characteristic feature of earlier discussions of 4Dism in the twentieth century is the conflation of the problem of persistence with the question of the ontological status of past and future entities and moments of time—the central problem in the philosophy of time. Although these issues are obviously related, they are distinct and, according to most authors, one's view of time does not entail a particular view of persistence.[27] This is not to blame the earlier authors for any oversight but to note that in reading their work, one should be careful not to mistake an argument in favor of "manifold realism" (see §1.1.3) for an argument supporting a 4Dist account of persistence.

Let us briefly examine some representative writings of Quine, in which he brings relativistic considerations to bear on both issues.

Just as forward and backward are distinguishable only relative to an orientation, so, according to Einstein's relativity principle, space and time are distinguishable only relative to a velocity. This discovery leaves no reasonable alternative to treating time as spacetime.   (Quine 1960: 172)

The difference between space and time thus ceases to be absolute and becomes relative to the circumstances of the observer, like the difference between left and straight ahead in the case of the stakes. The temporal dimension must thenceforward be fully integrated with the three spatial ones. We must accommodate all sorts of spatiotemporal diagonals. If a flash is to occur a mile north of here an hour hence, we must be able to say how "far" the flash is from here-now in absolute space-time units, valid for all observers, along a direct spatiotemporal diagonal. I say "we must"; of course it is all done and in the books. It runs into imaginary numbers and the speed of light in ways that happily need not detain us.   (Quine 1987: 198–9)

---

[24] The *locus classicus* of skepticism about strict identity across time is Hume [1739] 1978: 251ff. The *locus classicus* of skepticism about identification across time is Quine 1963.

[25] See, in particular, Quine 1953: 442 f and 1960: 170–6.

[26] Richard Taylor's classical work (1955), which explores such analogies, presupposes the doctrine of temporal parts without making this controversial presupposition explicit.

[27] For recent discussions of this topic, see Carter and Hestevold 1994; Merricks 1995; Lombard 1999; Sider 2001: §3.4; Haslanger 2003; Sattig 2006; and §1.1 of this book.

These passages convey the idea that since in SR space and time do not stand separately but form a union, which alone enjoys the invariant status, one should treat space and time on a par and, hence, regard "temporally distant" entities and locations on a par with spatially distant entities and locations. Reality must be four-dimensional.

Does it mean that physical objects must themselves be four-dimensional and persist by having distinct temporal parts at different regions of spacetime? Quine seems to think it does:

When time is thus viewed [i.e., as an integral part of the 4D spacetime manifold], an enduring solid is seen as spreading out in four dimensions: (1) up and down, (2) right and left, (3) forward and backward, (4) hence and ago. Change is not thereby repudiated in favor of eternal static reality, as some have supposed. Change is still there, with all its fresh surprises. It is merely incorporated. To speak of a body as changing is to say that its later stages differ from its earlier stages, just as its upper parts differ from its lower parts. Its later shape need be no more readily inferred from its earlier shape than its upper shape from its lower.   (Quine 1987: 197)

[T]he space-time view helps one appreciate that there is no reason why my first and fifth decades should not, like my head and feet, count as parts of the same man, however dissimilar. . . . Physical objects, conceived thus four-dimensionally in space-time, are not to be distinguished from events or, in the concrete sense, processes.   (Quine 1960: 171)

These are surely very explicit statements of the doctrine of temporal parts. But one is not forced to uphold this doctrine simply in virtue of the fact that in SR, space and time become, in a relevant sense, "interrelated." To resist this further step one could simply note that persisting objects are capable of multilocation in spacetime. Hence the selfsame enduring object could "comprise the content" of many distinct achronal regions of spacetime. Such an object would have "spatial" (i.e., achronal) parts but no "temporal" (i.e. diachronic) parts. The fact that what is, from one standpoint, "merely spatial" includes both "spatial" and "temporal" in another perspective in Minkowski spacetime does not go as far as obliterating the difference between achronal and diachronic parts. They still stand separately: no two distinct achronal parts of o at an achronal region R can be its distinct diachronic parts at two different achronal regions. In other words, no region can be both achronal and non-achronal.

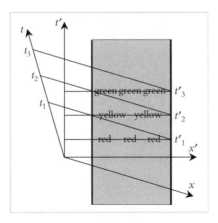

Figure 5.2. A stick changes its color. Adapted from Smart 1987: 64.

It is unclear whether Quine intended his considerations to serve as a forceful argument from SR to 4Dism about persistence or merely as a suggestive picture. Be that as it may, proceeding from somewhat different observations, J. J. C. Smart did offer a sketch of such an argument in favor of the "Minkowskian view of physical objects" as extended in time and persisting by having temporal slices and against the "Strawsono-Aristotelian view of objects" as enduring entities (Smart [1972], reprinted in Smart 1987):

> We have noted that an SA [Strawsono-Aristotelian] object is not (say) spherical *tout court*, but is spherical *at a time*. Thus the SA scheme is tied to the notion of a present instant at which the whole SA object possesses certain properties. Since the advent of special relativity, however, it has become clear that there is no absolute simultaneity. Two parts of an extensive SA object might, for example, simultaneously have the same colour with respect to one set of axes and have different colours with respect to a different set of axes.
>
> . . .
>
> So with respect to what axes must the Strawsono-Aristotelian think of an object as being "all at once"? A natural answer might be: a set of axes with respect to which the object is at rest. This would appear at first sight to remove ambiguity. However, what about an extensive Strawsono-Aristotelian object rotating rapidly on its axis? And what about a complex of SA objects all moving at different velocities relative to one another?   (Smart 1987: 63–4)

The situation envisaged by Smart is represented in Figure 5.2. For simplicity, let us assume that the object is a stick that starts out red-all-along at t'_1 in its rest frame, abruptly changes its color to yellow at t'_2 and then to green at t'_3, and let us abstract from o's earlier and later career. The argument then appears to go as follows:

1. An extended enduring object o in Minkowski spacetime is red-all-along at a certain time in one reference frame and red-yellow at a certain time in another reference frame.
2. No reference frame is objectively privileged in SR.
3. o is neither unambiguously uniformly colored at some time nor unambiguously not-uniformly colored at all times.
4. No object can be neither unambiguously Φ at some time in its career nor unambiguously $\neg\Phi$ at all times in its career.
5. If the spacetime of our world is Minkowskian then objects do not endure.

When the argument is put in this form, it becomes clear that step 4 is ambiguous. It could be read as asserting that no object can both have a certain property at some moment or other of its career and lack that property at all moments of its career, in a *particular* frame of reference. When read this way, the statement appears to be true but the argument becomes invalid for, according to the scenario, there is no single frame in which o is both uniformly colored at some time and not uniformly colored at all times. On the other hand, (4) could be read as asserting that no object can both have a certain property at some moment of its career in *some* frame and lack this property at all moments of its career in some—perhaps a *different*—frame. Then the scenario presents a genuine counterexample to this claim, but the claim itself loses all its force. As noted above (§5.1), properties in special relativistic spacetime must be relativized (according to one of the available schemes), not merely to moments of time, but to moments of time in a particular frame of reference. While no object can be Φ at some time in F and $\neg\Phi$ at all times in F, there is nothing problematic in being Φ at some time in F_1 and $\neg\Phi$ at all times in F_2.

Smart's concern[28] seems to be that extended enduring objects may have achronal parts with incompatible color patterns, depending on which

---

[28] Recently seconded by Petkov 2005: 136–9.

reference frame they are considered to be "all at once." But given that enduring objects are capable of multilocation and can reside, "all at once," in *all* legitimate reference frames, this concern really amounts to a restatement of endurantism in the framework of SR. A single enduring object can be located at multiple (including "crisscrossing") flat achronal regions of Minkowski spacetime and at all such regions it must have a determinate property. This is quite consistent with having contradictory properties at crisscrossing regions.

To sum up, while Quine's and Smart's analyses undoubtedly demonstrate the advantages of thinking of objects as extended in the diachronic dimension in Minkowski spacetime, and thus persisting by having diachronic parts, they stop short of providing convincing arguments. It should also be mentioned that what Quine and Smart say in the above-cited works suggests the beginning of a different argument from SR to perdurantism having to do with spatially extended objects. This argument will be developed and defended in Chapter 8.

5.4. "Profligate Ontology"?

Our brief review of Quine's and Smart's early relativistic considerations against endurantism highlights the importance of a charitable understanding of this doctrine in the new spacetime framework. In making a transition from the Galilean to Minkowski spacetime, everyone—endurantists, exdurantists, and perdurantists alike—should be allowed to make adjustments. To be sure, the adjustments must preserve the spirit of these views. On the other hand, if the adjustments are not made, the resulting doctrines may fall prey, all too quickly, to easy but irrelevant objections. The present section illustrates this point further.

In a recent paper, Steven Hales and Timothy Johnson have made two related claims: (i) endurantism, as ordinarily understood, is inadequate in the context of SR; (ii) although it could be modified so as to be consistent with SR, "the result is a profligate, Byzantine ontology with few redeeming qualities" (Hales and Johnson 2003: 525). A notable feature of Hales and Johnson's approach is that they do not wish to rely on the spacetime geometrical interpretation of SR. They believe that their conclusions are supported solely by the two postulates of SR (i.e. that physical laws are the same in all inertial frames of reference and that the speed of light in a vacuum is a constant) and the main consequence of the postulates, the relativity of simultaneity.[29]

Hales and Johnson's observations have to do with spatially (i.e. achronally) extended objects and turn on the central thesis of endurantism, that enduring objects are wholly present at all moments of time at which they exist. Consider an enduring stick and two moments of time in its rest frame F', t'_1 and t'_2. Let A_1 and B_1 be two small material parts at the ends of the stick at t'_1, and A_2 and B_2 two such (perhaps distinct, if the stick underwent mereological change between t'_1 and t'_2) parts of it at t'_2 (see Figure 5.3).[30]

1. A_1 and B_1 are simultaneous in F'.
2. Simultaneity is sufficient for coexistence.
3. A_1 and B_1 coexist. (1, 2)
4. A_1 and B_2 are simultaneous in F (a different inertial frame of reference).
5. A_1 and B_1 coexist; A_1 and B_2 coexist. (2, 3, 4)
6. Coexistence is transitive.
7. B_1 and B_2 coexist. (5, 6)

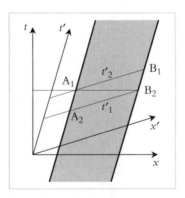

Figure 5.3. Enduring stick.

[29] Hales and Johnson also wish to abstract from the question of whether eternalism or presentism is true (ibid.: 528). It is unclear how presentism could be true in the special relativistic context, and Hales and Johnson do not elaborate. Indeed, they note later on that the relativity of simultaneity "seems to drive a stake through the heart of presentism" (ibid.: 533).

[30] Hales and Johnson's presentation does not make it clear that A_1, A_2, B_1, and B_2 are material parts of the stick, as opposed to the spacetime points they occupy. I hope my reconstruction is true to their original intention.

8. If x and y are coexisting material parts of an enduring object, it is not wholly present at a single time unless x and y exist at one time.

9. The stick is not wholly present at a single time because it has coexisting parts that do not exist at the same time in F': namely, B_1 and B_2, and also A_1 and B_2. (5, 7, 8)

Hales and Johnson conclude that since the stick is not wholly present at a single time, it cannot be an enduring object. In contrast, no problem arises if the stick is a perduring object, for no chosen collection of temporal parts of such an object need be wholly present at any single time.

The argument is clearly invalid.[31] Note, first of all, that the crucial premise (8) leaves out any mention of a frame of reference in which x and y both exist at a certain time. This is good, for no fixed frame would do the job Hales and Johnson expect from (8). Rather the force of (8) is:

8'. If x and y are coexisting material parts of an enduring object, it is not wholly present at a single time unless x and y exist at one time *in some inertial frame of reference or other*.

When (8) is thus refined the fact that A_1 and B_2 do not exist at any one time in F' does not prevent them from existing at a single time in some other frame (namely, F). This disqualifies A_1 and B_2 from creating a problem for the presence of an enduring stick in its entirety at a single moment of time, in F. The problem still appears to arise for B_1 and B_2, for they clearly belong to different moments of time in *any* frame. Here, however, it should be noted that the transitive relation of coexistence, which leads, as per (7), to the problem, is not appropriate for (8). The notion of coexistence, according to which physical objects such as B_1 and B_2, located as they are at timelike separated spacetime regions (using the language of the standard geometrical interpretation of SR), can be said to coexist with each other should bother the endurantist no more than the trivial relation of coexistence holding between any two regions of a generic spacetime manifold. All such regions and their occupants trivially coexist (in the appropriate tenseless sense) with each other, but this is quite consistent with the more interesting fact about an enduring object, namely, that such an object is wholly present, or located, *only* at *achronal* subregions

---

[31] There are several flaws in this argument and the following is one way to expose some of them. Cf. Miller 2004 and Gibson and Pooley 2006: 177–9 for similar criticisms.

of the manifold—that is to say, at many o-eligible achronal slices of its path (see §5.1).

Hales and Johnson are aware of it but think that conceding the latter statement would lead to a "metaphysically unsettling profligate multiplicity" of objects. Instead of one enduring stick, we would have two: A_1-B_1 as well as A_1-B_2, and once we had two, we would have an infinity of others, occupying all (o-eligible) achronal slices of the stick's path. But this claim is in error. There is only one enduring stick. The fact that it is (as manifold eternalists would put it) capable of multilocation in spacetime and can exhibit different properties and material composition at different achronal slices of its path—or (to put it in the form friendly to Hales and Johnson's intention to do without invoking the notion of spacetime) at different moments of time in different frames of reference—does not make a single object many; not any more than relativizing properties and spatial composition to moments of absolute time multiplies enduring objects in the classical framework.

What this fact does show is that, rather than being wholly present simply at "moments of time," enduring objects are wholly present at appropriate achronal regions (that could be indexed by time-frame pairs, see §5.1), and their properties and achronal parthood must be relativized accordingly. Failure to relativize them to frames, as well as times, imposes on the endurantist an inconsistent combination of relativistic and classical ideas: relativistic, insofar as the kinematical effects of SR, including the relativity of simultaneity, are accepted; classical, insofar as possession of properties and achronal parts is not similarly relativized. Surely, this combination lands one in trouble. But why should the relativistically informed endurantist subscribe to such a combination? To repeat, it would be quite similar to unwarranted refusal to relativize properties to moments of absolute time in the classical framework.

5.5. Is Achronal Universalism Tenable in Minkowski Spacetime?

The above-noted unsuccessful attempts to refute endurantism in the context of SR are based, in essence, on imposing on the endurantist an unwarranted restriction by disallowing her to talk of enduring objects being located at *all* object-eligible (flat) achronal slices through their paths. My response on behalf of the relativistically minded endurantist was, in essence, to point out that preventing her from speaking of the location of enduring objects at multiple t^F-slices of their paths in Minkowski spacetime (and thus from the corresponding relativization of properties and achronal composition) is no more justified than preventing the classical endurantist from speaking of the location of enduring objects at multiple t-slices of their paths in Galilean spacetime.

But this, by itself, does not tell us *what* t^F-slices of o's path are o-eligible. The question presents particular interest when it is contrasted with a similar question that was earlier posed in the Galilean framework: *what t-slices of o's path are o-eligible?* There the answer was straightforward: *all* of them, the thesis we called Galilean Achronal Universalism (§4.1). Does a similar thesis hold in Minkowski spacetime:

> (AU^M) (i) Any enduring object is located at *every* t^F-slice of its path (in Minkowski spacetime), (ii) any perduring object has a t^F-part at every t^F-slice of its path, and (iii) any exduring object is q-located at every t^F-slice of its path?

In other words, is every t^F-slice of o's path o-eligible? The Unicolor example (§5.2) has already raised some doubts pointing to their underlying source, the ubiquitous presence of *crisscrossing* t^F-slices in special relativistic spacetime. The problem is particularly acute for endurance and exdurance theories, and less pressing for the perdurance theory. In the context of endurantism and exdurantism, the question of *what t^F-slices of o's path are o-eligible* becomes the question of *where* in spacetime the objects *themselves* are located (or q-located). *This* question does not present any special problem for perdurantism. A perduring object is located at its path and nowhere else. There is an issue in the neighborhood: what t^F-slices of a perduring object's path are eligible to host its diachronic parts? But there is no corresponding issue about the object itself.

In a recent discussion of the problem of o-eligibility[32] Cody Gilmore (2006) has argued, in effect, that clause (i) of (AU^M) must be rejected by anyone who endorses the existence of composite objects and the standard spacetime interpretation of the relativity theory. The argument

[32] But not under this name.

is rather complicated and is part of a longer attack on the conjunction of the endurantist thesis of multilocation with the latter two views, the consequence being that accepting the standard interpretation of relativity and the existence of composite objects raises serious problems for *locational endurantism*.[33]

Below I focus on the part specifically having to do with (AU^M) (Gilmore's name for it is "Every Slice Principle"). That part itself has three sections developing distinct arguments. I shall not deal with one of the sections, which considers impenetrable extended simples and time travel in curved spacetime of GR, and this for three reasons: (i) except for a few brief comments here and there, the present study does not discuss general relativistic spacetime; (ii) the present study abstracts from the possibility of extended simples (see §1.2); and (iii) Gilmore's argument against the GR version of Achronal Universalism, based on a scenario involving extended simples in curved spacetime, was recently criticized by Gibson and Pooley (2006: 184).[34] Accordingly, I shall focus on the other two arguments.[35] In difference from Gilmore, however, my discussion will cover both endurance and exdurance. For the most part, this is done by replacing 'location' (or its analogs) with 'q-location.'

Both arguments have to do with crisscrossing achronal slices of objects' paths in Minkowski spacetime. The first raises a problem for the sort of "unbridled crisscrossing" that would be sanctioned by clauses (i) and (iii) of (AU^M). Gilmore, however, takes this to compromise, not merely (AU^M), but the idea of crisscrossing in general, the claim further supported with another argument.

In the remainder of this section I review Gilmore's first argument (from "corner slices") but take its lesson to be more limited: while it shows (AU^M) to be untenable, it does little to undermine "disciplined crisscrossing" within the limits of eligibility. In the next section I proceed

---

[33] That is, the view that objects persist by being multilocated at achronal slices of their paths in spacetime. On Gilmore's fourfold classification of the modes of persistence (see §2.5), locational endurance is consistent with mereological perdurance (i.e. the having of diachronic parts). Gilmore revisits his strategy in §6 of Gilmore 2008.

[34] But they offer what they consider to be an improved version of that argument on pages 185–6.

[35] I recast Gilmore's arguments in my own terminology throughout the ensuing discussion. This, by itself, should not create any problems, for despite some disagreement (e.g. about exdurance as a legitimate mode of persistence and about the importance of certain remote possibilities, see §2.4.6), my general approach to ways of defining various modes of persistence in spacetime is close to his.

to defend "disciplined crisscrossing" against Gilmore's second argument (from "immanent causation").

Consider a complete path of an enduring or exduring object o composed of four atoms o_1, o_2, o_3 and o_4 in Minkowski spacetime (Figure 5.4):

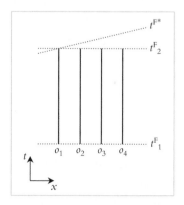

Figure 5.4. "Corner slice" (Gilmore 2006: 212–13).

o_1, o_2, o_3 and o_4 pop into existence at t^F_1 and go out of existence at t^F_2. Both t^F_1-and t^F_2-slices of o's path o are good candidates for o's q-location, and so are all the t^F-slices for any $t^F \in (t^F_1; t^F_2)$. Consider, however, the t^{F*}-slice in frame F* distinct from F. It does count as an achronal slice of o (see D5.4^M and D5.8^M of §5.1). According to AU^M, o must be q-located in it. But this is problematic, for the t^{F*}-slice of o is a "corner slice" that contains a single atom o_1 and can hardly qualify for being a suitable q-location for the entire object o. Recall that on our understanding of the basic notion of (q-)location, a region at which an object is exactly (q-)located, or "wholly present," is the region into which the object exactly fits and which has exactly the same size, shape, and position as the object itself (see §2.4.3). But the t^{F*}-slice of o is shaped like a single atom and, hence, not shaped like o. An object that is, for most of its career, composed of four atoms cannot "fit into" a region shaped like one atom. t^{F*}-slice of o is clearly not o-eligible. This refutes AU^M.

The friend of AU^M may attempt to reply to the argument from corner slices by rejecting the possibility of scenarios such as one described above. The careers of all the objects represented in it violate the conservation laws of physics (because the careers represent objects as popping into and out

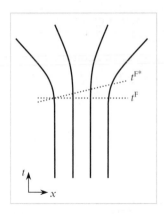

Figure 5.5. A composite object "breaks up" at t^F.

of existence), whereas the relativistic considerations underlying this whole discussion and motivating AU^M in the first place presuppose their validity. This makes physically impossible states of affairs, such as that depicted in Figure 5.4, irrelevant to the discussion at hand, even if they are not impossible *tout court*.

We shall return to this general point about the dialectic of the dispute at the end of §5.6. For now, it will suffice to note that the particular concern it raises for the corner slice scenario can be alleviated by a harmless modification (suggested by Gilmore in personal correspondence). Suppose that, instead of popping in and out of existence, the atomic parts of o suddenly "break up" (i.e. cease to compose o) at t^F and begin to separate (see Figure 5.5). What do we now say of the t^{F^*}-slice of o's path o? It still appears to contain a single atom, so the problem recurs, for on Achronal Universalism^M, o must, impossibly, be q-located at a single spacetime point.[36] But conservation laws are now respected.

One way to respond to this move might be by questioning the notion of instantaneous "break-up" of o at t^F. One wants to know more about the conditions in which o ceases to exist due to mutual separation of its atoms. In all likelihood, the conditions will be vague, resulting in an extended interval of "fading away," with no sharp temporal boundaries, such as t^F. Hence it is not so clear, after all, that the t^{F^*}-slice of o is not o-eligible.

---

[36] One has to be careful here, for mutual spatial separation of the atoms at t^{F^*} could, in principle, be offset by relativistic length contraction, thus making t^{F^*} o-eligible! However, even if t^{F^*} is not fit for the job, another suitable flat slice, perhaps in another frame of reference, can always be found.

This may depend, in a complicated way, on the fine details of the relevant theory of spatial composition, the nature of the object in question, and the exact trajectories of its atoms. And even when all that is taken into account, the answer will perhaps remain vague.

But upon reflection, considerations of this sort cannot save AU^M. On any view of vagueness, some t^F-slice of o or other will not be o-eligible (perhaps, on some precisification), or at least not determinately o-eligible, and thus in conflict with AU^M. What the above considerations do show is that questions of o-eligibility are logically *prior* to questions about the exact boundaries of o's path in relativistic spacetime: they need to be settled first.[37] While this compromises AU^M (which mistakenly attempts to put the cart before the horse), it does not suggest that anything is wrong with object-eligible crisscrossing slices in general. It merely suggests that "unbridled crisscrossing" must be rejected in favor of "disciplined crisscrossing."

Notice that this leaves the relevant "disciplinary constraints" outside the scope of this discussion. But this is as it should be, and the situation here is no more (even if no less) problematic than it was in Galilean spacetime. To see this, return to Figure 5.4 and consider the evolution of o in F*. From the physical point of view, F* is a legitimate frame of reference, which describes o as progressively losing parts, one by one. How many atomic parts could o lose without ceasing to exist? Maybe just a few, or maybe the majority of them. Perhaps there is no general answer and it all depends on the nature of the object in question. But when the evolution of o is viewed from a perspective in which it looks gradual, it becomes clear that (i) questions of this sort must indeed be settled *before* one attempts to draw the exact boundaries of o's path, and (ii) *exactly the same* questions would arise if spacetime were classical and t^{F*}-planes represented the absolute time planes.

It should be noted that in Gilmore (2006: 212–13), Gilmore himself regards the case of corner slices not so much as a problem for Achronal Universalism in general, but as an occasion to modify a particular version of it—namely, AU^M—by (i) abandoning what I earlier dubbed FLAT (and thus allowing persisting objects to be located at arbitrary, not necessarily flat, achronal slices of their paths) and (ii) imposing a *maximality* condition on o-eligibility. The latter amounts to refusing o-eligibility to any achronal slice of o's path, unless the slice in question is not a proper subregion of a larger

---

[37] Cf. Gibson and Pooley 2006: 186–7, who make a very similar suggestion.

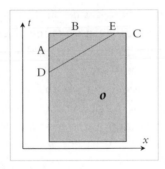

Figure 5.6. Maximal achronal slices.

o-eligible achronal slice of o's path. With these modifications, no corner slice of o (e.g. AB in Figure 5.6) is o-eligible, because such a slice is a proper subregion of a larger o-eligible non-flat achronal slice of o (namely, ABC). Given my adherence to FLAT, I must reject this move.[38] Quite apart from that, Gibson and Pooley rightly note (2006: 186) that Gilmore's suggested modifications to AU^M may be too restrictive, as they automatically refuse o-eligibility to all less-than-maximal achronal slices (such as DE in Figure 5.6) of o's path. This leads to problematic verdicts in some cases. Suppose a suitably more fine-grained version of DE represents the q-location of my body particles, except for one of my hairs. On the modified version of Achronal Universalism, I am not q-located at DE; surely a strange result.

To sum up this section, Achronal Universalism does not survive the transition from Galilean to Minkowski spacetime, the reason being the widespread presence of "unacceptable crisscrossing" (as in Figures 5.4 and 5.5). The problem so far seems to lie with "unacceptable," not with "crisscrossing." But is any crisscrossing acceptable? Gilmore suggests not (2006: 214–19 and 2008: 1244–6). I now turn to his second argument.

5.6. "Crisscrossing" and Immanent Causation

Immanent causation is a broadly causal relation between an earlier and a later state of the same object.[39] If you stain this book by putting a coffee cup

---

[38] For discussion and defense of (FLAT) see §5.2 above.
[39] See §§4.4–4.6, where immanent causation was employed in an objection to the argument from vagueness to 4Dism.

on it, it will remain stained at later times. Despite its ubiquity, immanent causation is not easy to make precise. But then the same is true even of the ordinary causal relation between the states of different entities. What matters for our purposes is that the following principle dubbed by Gilmore MURIC (2006: 214), for "Multilocation Requires Immanent Causation," is highly plausible:[40]

> (MURIC) If an enduring (or exduring) object is q-located at two spacetime regions, the contents of these regions must be related by immanent causation.

Suppose the contents of two distinct spacetime regions R_1 and R_2 do not stand in any such relation, that is to say, the properties of the contents of R_2 do not depend, in a relevant sense to be made precise by a general theory of immanent causation, on the properties of the contents of R_1, and vice versa. Then there is a strong reason to conclude that no single object is exactly q-located at both R_1 and R_2. For if it were then what happened to the object at one of these regions would leave its mark on its state at the other region.

To illustrate this idea further, consider two contrasts. One is between R_1 and R_2 in Figure 5.7(a), on the one hand, and R_1^* and R_2^* in Figure 5.7(b), on the other.[41] R_1 and R_2 could well host the same enduring or exduring object, but R_1^* and R_2^* could not: "too much" of R_1^* is spacelike separated from "too much" of R_2^* and, therefore, "too much"

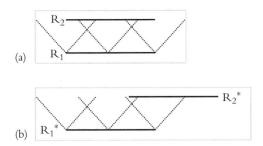

Figure 5.7. The contents of R_1 and R_2 are (a), but the contents of R_1^* and R_2^* are not (b), causally related.

---

[40] My statement of MURIC differs from Gilmore's in that it accommodates both location and q-location and thus extends his considerations to exduring as well as enduring objects.

[41] Figures 5.7(a) and 5.7(b) are pale renditions of Figures 44 and 46 in Gilmore 2004: 294, 296.

of the contents of $R_1{}^*$ is causally isolated from "too much" of the contents of $R_2{}^*$. (The dotted lines represent lightcones.) There is still a connection between the contents of $R_1{}^*$ and $R_2{}^*$, but it is too tenuous to cement their strict identity (in the case of enduring objects) or genidentity (in the case of exduring objects).

Consider also the contrast between any two achronal slices of the path of a single pointlike enduring or exduring object, on the one hand, and two analogous slices which, however, belong to distinct but qualitatively identical objects involved in an "immaculate-replacement" scenario.[42] In the first case, the contents of the slices (i.e. of single points in spacetime) stand in an appropriate relation of immanent causation, while in the second, they do not—just as it should be. MURIC gives correct verdicts in both cases.

Thus MURIC has a lot to recommend it. But Gilmore argues that it presents a problem for crisscrossing slices of objects' paths in Minkowski spacetime. The problem arises as follows. Consider two crisscrossing achronal slices $o_{\perp 1}$ and $o_{\perp 2}$ of the path of some achronally extended enduring or exduring object o (i.e. an object having proper spatial parts; Figure 5.8):

We can safely assume that both $o_{\perp 1}$ and $o_{\perp 2}$ are o-eligible (neither of them is a "corner slice," etc.). But their contents are not related by

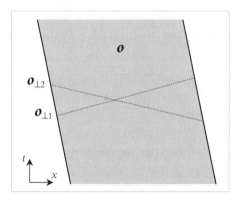

Figure 5.8. Crisscrossing slices of an achronally extended object (Gilmore 2006: 215).

---

[42] "Immaculate replacement" is an event in which one object is instantaneously annihilated and another qualitatively identical object is simultaneously created at an "immediately succeeding" spacetime point or region. The term 'immaculate replacement' was coined by Swoyer 1984: 598.

immanent causation. The reason is that "too much" of $o_{\perp 1}$ is spacelike separated from "too much" of $o_{\perp 2}$ and, therefore, "too much" of the contents of $o_{\perp 1}$ is causally isolated from "too much" of the contents of $o_{\perp 2}$, and vice versa—just like in the case of R_1^* and R_2^* above (Figure 5.7(b)). In accordance with MURIC, then, $o_{\perp 1}$ and $o_{\perp 2}$ cannot host the same enduring or exduring object. This unwelcome result is in conflict with the presumed o-eligibility of $o_{\perp 1}$ and $o_{\perp 2}$.

One radical way to block this result is to eliminate the possibility of crisscrossing slices by denying the existence of achronally extended objects. That would rule out composite objects and thus amount to adopting a version of compositional nihilism.[43] Few of us would be willing to go that far. But Gilmore offers a different remedy: to attempt modifying MURIC in a way that would still allow it to give a correct verdict in "immaculate replacement" scenarios but would not revoke o-eligibility from crisscrossing slices, such as those in Figure 5.8. He considers the following candidate for replacing MURIC:[44]

(MURIC*)   If an enduring or exduring object is q-located at two spacetime regions, then no matter whether or not the contents of these regions *themselves* are related by immanent causation, it must at least be the case that these regions are slices through the path o of o, such that o can be partitioned into a set of slices $\{o_{\perp}(\tau), \tau \in T\}$, such that the appropriate sort of immanent causal relation holds between the contents of $o_{\perp}(\tau_1)$ and $o_{\perp}(\tau_2)$, for any $\tau_1, \tau_2 \in T$, $\tau_1 \neq \tau_2$.

Here τ is some useful real-valued ordering parameter. In accordance with FLAT, it is assumed that all the relevant slices of o are flat.[45]

Unlike MURIC, MURIC* renders crisscrossing slices object-eligible. Return to the situation represented in Figure 5.8. Although the contents of two crisscrossing slices $o_{\perp 1}$ and $o_{\perp 2}$ of o's path o are not themselves

---

[43] Compositional nihilism is the view that no objects have proper achronal parts. Strictly speaking, this does not rule out the existence of achronally extended simples. But we abstract from the possibility of achronally or diachronically extended simples in this study (see §1.2).

[44] Gilmore 2006: 216. Again, I take the liberty to put MURIC* in my own terms and extend it to q-locations as well as location. See also n. 40.

[45] Gilmore himself does not make this assumption. But he does not explicitly reject it either. Although his consideration is thus not officially restricted to flat achronal slices of an object's path his discussion involves, for the most part, only flat slices. In difference from Gilmore, I treat the cases of flat and non-flat slices separately.

related by immanent causation, o can easily be partitioned into a set of slices whose contents are so related, for example, a family of t-slices in reference frame (x,t).

On the other hand, just like MURIC, MURIC* gives a correct verdict in the immaculate replacement (henceforth 'IR') scenario involving a pointlike object. The path of such an object is a one-dimensional timelike line which *cannot* be partitioned into a set of points, such that their contents are all pairwise immanent-causation related. Any such partition would feature a pair of points on the opposite sides of the "immaculate replacement gap" (henceforth 'IR-gap') which are not so related.

One would think that this consideration could easily be extended to IR scenarios involving achronally composite objects. But there is a problem here, brought to light by Gilmore (2006: 217–19). Suppose God annihilates a persisting object a, composed of an array of point particles, at a certain moment in some frame (x,t) and immediately creates a qualitatively identical array a'. For simplicity, suppose t_{last} is the last moment of existence of the original array in (x,t), but there is no first moment of existence of the newly created array; rather there are infinitely many "first moments" arbitrarily close to the IR-gap (see Figure 5.9).[46] Focus on one of such moments, t_{first}. There is, in these circumstances, strong pressure to say that the t_{last}-slice of a's path and the t_{first}-slice of a''s path cannot host the same persisting object. Indeed, it is pretty clear that they host distinct objects,

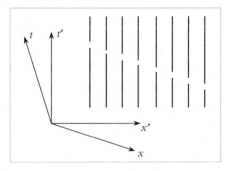

Figure 5.9. IR-gaps are simultaneous in frame (x,t), but not in frame (x',t') (Gilmore 2006: 217).

---

[46] In the situation envisaged there is, of course, no literal spatiotemporal gap. The term 'IR-gap' and the gap in the corresponding diagrams are intended to refer informally to the geometrical locus of the immaculate micro-replacements.

a and *a'*. But MURIC* fails to rule out a hypothesis to the contrary. How so?

Suppose that such a hypothesis is true, that is to say, a single persisting object *o* is q-located at both the t_{last}-slice of *a*'s path and the t_{first}-slice of *a'*'s path. The only reasonable candidate for the path of this object is the union *o* of the paths of *a* and *a'*: $o = a \cup a'$. To use MURIC* to rule out the existence of *o* would require showing that *o* *cannot* be partitioned into a set of slices $o_{\perp}(\tau)$, $\tau \in T$, such that the appropriate sort of immanent causal relation holds between the contents of $o_{\perp}(\tau_1)$ and $o_{\perp}(\tau_2)$, for any $\tau_1, \tau_2 \in T$, $\tau_1 \neq \tau_2$. But consider a partition of *o* into a family of t'-slices in some inertial reference frame (x',t') distinct from (x,t) (i.e. "horizontal" slices in Figure 5.9) and pay particular attention to those of them that intersect the IR-gap. The material contents of any two sufficiently close such slices $o_{\perp n}$ and $o_{\perp (n+1)}$ are either the same or differ, at most, in a single point particle. If we are willing to tolerate persistence through gain and loss of single particles then we should grant that the contents of $o_{\perp n}$ and $o_{\perp (n+1)}$ are appropriately related by immanent causation. Assuming that such a relation is transitive leads to the conclusion that any two t'-slices of *o* are so related. Accordingly, the t_{last}-slice of *a* and the t_{first}-slice of *a'* are, after all, slices through the path *o*, which *can* be partitioned into a set of slices, namely $\{o_{\perp}(t')\}$, such that the appropriate sort of immanent causal relation holds between the contents of $o_{\perp}(t'_1)$ and $o_{\perp}(t'_2)$, for any $t'_1 \neq t'_2$. This means that MURIC* fails to rule out the existence of a single object q-located at both $a_{\perp}(t_{\text{last}})$ and $a'_{\perp}(t_{\text{first}})$ and, hence, MURIC* is not a good candidate to replace the already compromised MURIC. In one sense, the latter was too restrictive. But its successor, MURIC*, is too permissive.

In addition to thus showing MURIC* to be overly permissive, one could also offer a general analysis of *why* it might be inferior in spirit to MURIC. While MURIC attempts (in the end, unsuccessfully) to ground identity or genidentity of a persisting object in what seems to be highly relevant to it—the immediate and open-to-easy-inspection relation of immanent causation between the contents of two spacetime regions—MURIC* takes an extremely roundabout route purporting to ground the relation between such contents in (what appears to be) a conceptually distant existential fact about the relation between the contents of some *other* regions leading, in the end, to a watered-down criterion.

Below I offer two replies to these qualms about MURIC*.

(i) Return, first, to the inference from the fact that the contents of any two sufficiently close slices $o_{\perp n}$ and $o_{\perp (n+1)}$ of o (the path of a single persisting object o hypothesized to exist) "at an angle" to the IR-gap—which differ, at most, in one point particle—are appropriately related by immanent causation to the fact that, because of the transitivity of immanent causation, any two t'-slices of o are so related. This inference involves a "Sorites series" and, perhaps, in an objectionable way. To set things in perspective, consider an analogous Theseus ship-style Sorites series of particle-by-particle mereological replacements of the material contents of an achronally extended enduring or exduring object leading, in the end, to a completely "renovated" object. One may be willing to tolerate persistence through gain and loss of a single particle, at each step of this process, and yet hesitant to apply the "principle of transitivity" to conclude that the object survives the entire Sorites series.[47] The point is not that one *must* reject the transitivity of persistence in such cases but only to note that employing a Sorites series in argumentation about persistence lands one in a generally problematic area and opens a familiar can of worms. It is now easy to see that the case of gradual "step-by-step" persistence across an achronally extended IR-gap (in reference frame (x',t') in which the t'-slices are "at an angle" to the gap) is in the same category.

(ii) Notwithstanding any Sorites-type considerations, one may be inclined to agree with Gilmore that MURIC* is indeed too weak to ban counter examples, such as one sporting an achronally extended IR-gap, but argue that a better candidate for replacing MURIC is available. I favor the following:[48]

[47] Cf. Gibson and Pooley 2006: 183, who make essentially the same point about Gilmore's problematic employment of the "principle of transitivity."

[48] My version of MURIC** is essentially similar to Gilmore's who considers and rejects it in an extended endnote in Gilmore 2006: 232–3, n. 42. His reasons for dissatisfaction with MURIC** are as follows. (a) MURIC** seems to be conceptually irrelevant to the issue of the causal relation between two given achronal slices of the hypothetical single object's path—in a way similar to that noted three paragraphs above in relation to MURIC*. (b) MURIC**'s negative verdict in the case of the achronally extended IR-gap is counterintuitive: it is not obvious that an object could not persist through the gradual immaculate replacement of its small parts. The issue of conceptual irrelevance is addressed immediately below. As regards (b), it has already been noted above that the appearance of counter-intuitiveness arises in such cases from the uncritical employment of the problematic "principle of transitivity."

(MURIC**) If an object o is q-located at two spacetime regions, then no matter whether or not the contents of these regions *themselves* are related by immanent causation, it must be the case that these regions are slices through the path o of o, such that for *any* partition of the relevant section of o into a set of t^F-slices in *any* inertial frame of reference F, $\{o_\perp(t^F), t^F \in T\}$, the appropriate sort of immanent causal relation holds between the contents of $o_\perp(t^F_1)$ and $o_\perp(t^F_2)$, for any $t^F_1, t^F_2 \in T$, $t^F_1 \neq t^F_2$.

Here T is the fragment of the entire lifespan of o in F corresponding to "the relevant section of o" including the two regions in question.

The crucial difference between MURIC* and MURIC** is as follows: while MURIC* grounds the appropriate relation between the contents of any two slices of the path o of a single persisting object in the existence of *at least one* partition (of a relevant section) of o into a set of slices whose contents all stand in a requisite broadly causal relation, MURIC** requires that *all* the partitions (of a relevant section) of o of a certain kind (namely, partitions into sets of t-slices in inertial frames of reference) feature immanent-causally related contents.

I submit that MURIC** is superior to MURIC. While relativistic physics gives strong reasons to think that, as required by MURIC**, there must be an immanent causal relation between the *successive* phases of a single object—i.e. the phases representing its state at successive moments of time in *any* arbitrary single inertial frame—there appears, *pace* MURIC, to be little or no reason to think that there must be the *same kind* of relation among the members of a haphazard collection of phases representing the states of the object at different moments of time in *different* reference frames. Physical laws in the relativistic framework (e.g. the Lorentz-invariant laws of motion) relate quantities (e.g. the 4-velocity, 4-acceleration, relativistic force, etc.) defined in a particular reference frame and can be used to infer a later state of an object from its earlier state, both states being defined in the same frame. Of course, as a result, the states defined in *distinct* frames also turn out to be related; but not by the *same* kind of a fundamental causal relation pertaining to the evolution of the object states in a given reference frame. Rather such states are related in a *derivative* manner, and sometimes—for example, in the case of crisscrossing slices—somewhat in a manner in which two effects of the same cause may be related.

MURIC** does justice to these considerations, while MURIC does not. In that respect, MURIC** scores well in terms of "conceptual adequacy" and thus addresses the "conceptual remoteness" concerns.

More importantly, MURIC** bans counterexamples of the sort described above, which involve achronally extended IR-gaps (see Figure 5.9). On MURIC**, to rule out the existence of a single enduring or exduring object o q-located at both the t_{last}-slice of a's path and the t_{first}-slice of a''s path requires showing that there is *at least one* partition (of the relevant section) of the purported path o of o sporting two slices that are *not* related by immanent causation. Such a partition is, of course, the partition of (the relevant section of) o into a family of t-slices, and the offending slices are none other than $a_\perp(t_{last})$ and $a'_\perp(t_{first})$.

It is also easy to see that MURIC** handles the case of crisscrossing slices just as well as MURIC*: any partition of (the relevant section of) the path of an object featuring two such slices into a set of t^F-slices in an arbitrary inertial frame of reference F displays a causally connected (in a relevant respect) sequence.

There is, however, a complication still to be considered. The above "immaculate replacement" scenario involves a *flat spacelike* IR-gap.[49] This is a natural and appropriate choice in the context of SR. The idea is that a spacelike hyperplane drawn through the spacetime points, at which the particles composing o are q-located at the last moments of their existence, is a good candidate to represent, in a certain inertial reference frame, the last moment of existence of the *entire* object. But there is certainly nothing necessary about this situation. The points at which the individual immaculate micro-replacements occur could, in principle, be distributed in spacetime in such a way that any hypersurface drawn through them would be curved or "corrugated" and at least partially timelike or lightlike.

Consider, first, the spacelike versus timelike (or lightlike) aspect of IR-gaps and abstract for the moment from the issue of curvature or

---

[49] As already mentioned (n. 45), Gilmore's discussion is not officially restricted to flat achronal slices, even though his explicitly developed examples involve only them. Since I accept FLAT (§5.2), I have very different attitudes towards flat and non-flat achronal regions in Minkowski spacetime. Hence my separate consideration of flat and non-flat slices in the context of discussing MURIC and its successors.

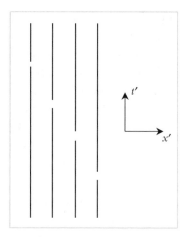

Figure 5.10. Timelike IR-gap.

"corrugation." I shall examine three types of "almost everywhere flat" hypersurfaces[50] drawn through the spacetime points at which the enduring or exduring particle constituents of a are q-located at the last moments of their existence: (i) everywhere spacelike, (ii) everywhere timelike, and (iii) "mixed" (partly spacelike, partly timelike).

Case (i) and its implications have already been discussed. Case (ii) (see Figure 5.10) depicts a situation in which the individual micro-replacements are so much separated "in time" that the whole process takes "very long" to complete.

There is, in this case, no frame of reference in which the macro-replacement happens "all at once." Indeed, there is nothing even remotely close. Rather the whole process becomes somewhat similar to the process of "slow and gradual" replacement of the parts of Theseus's ship. I submit that this situation cannot be analyzed as a case of an "immaculate replacement" of a, for the whole point of the latter is an *instantaneous* replacement of a with its qualitative duplicate a'. Case (ii), therefore, cannot serve as a basis of any objection to MURIC* (or MURIC**, for that matter).[51]

---

[50] The qualification "almost" is needed to accommodate the "mixed" case (iii).

[51] How *should* case (ii) be analyzed then? I do not have a firm opinion on the matter, but I suspect that anyone who has reasons to believe that the "replacement ship" in the Theseus ship scenario is identical (or genidentical) with the original ship may have equally good reasons to believe that, in the case of a timelike IR-gap, a' is identical (or genidential) with a.

Consider now a "mixed" case (iii), in which the "left half" of *a* undergoes an instantaneous immaculate replacement whereas its "right half" undergoes a "slow and gradual" Theseus ship-style replacement. What are we to say of the entire object *a*? In my opinion, it is utterly unclear what to say about *a* as a whole. The most one can say about the whole scenario is that it involves two objects (namely, the left and right halves of *a*) undergoing two very different sorts of processes, which must be analyzed along the lines of cases (i) and (ii) discussed above. Case (iii), therefore, does not raise any new problems for MURIC* (or MURIC**).

To be sure, cases (i)–(iii) do not exhaust all the possibilities. I contend, however, that the above brief examination of these cases strongly suggests that immaculate replacement scenarios involving partially or fully non-spacelike gaps have no relevance to the discussion of MURIC and its variants.

Turning now to the issue of curvature, an interesting case is an IR-gap that is everywhere spacelike but "nowhere flat" (see Figure 5.11). On the face of it, it presents a problem for MURIC**. Indeed, no partition of (the relevant section of) the union of *a* and *a'* into t^F-slices in any inertial reference frame features more than "just a few" micro-IR-gaps in the evolution of the alleged single object *o*. This may be reason enough to conclude that the sequences of such slices in all frames are causally connected.[52] And yet one might feel that the causal connection between the states of *a* and *a'* is lacking, for they are separated by an everywhere-spacelike, even if curved (or "highly corrugated"), macro-IR-gap.

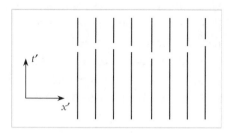

Figure 5.11. Non-flat spacelike IR-gap.

---

[52] For the sake of discussion, we put aside Sorites-related considerations, which could be adduced at this point in defense of MURIC** by analogy with similar such considerations used earlier to defend MURIC*.

Where would such a feeling come from? Not from any considerations based *solely* on special relativity. Curved and "corrugated" spacelike hypersurfaces lack a clear physical meaning in SR. To be sure, there is nothing illegitimate about them;[53] but unlike flat spacelike hypersurfaces, they do not represent any physically significant notion in this theory. The feeling that the careers of two persisting objects separated by a non–flat spacelike macro–IR–gap are not causally connected cannot, therefore, be based on considerations of relativistic dynamics in the way the corresponding conclusions about flat macro–IR–gaps are so based. If one still thinks that the feeling must be allowed to carry weight there must, I submit, be no good reason to disallow further modification in MURIC** intended to accommodate partitions of the path of a persisting object into families of spacelike hypersurfaces of arbitrary curvature.[54]

To elaborate, let us introduce the notion of an *S-partition* of spacetime region R in Minkowski spacetime:

(S-partition)   The set $\{S(\tau), \tau \in T\}$ of spacetime regions is an *S-partition* of region R in $ST^M =_{df}$ (i) for any τ, $S(\tau) \subseteq R$; (ii) any point in R belongs to some $S(\tau)$; (iii) $S(\tau)$ is everywhere spacelike; (iv) $S(\tau_1) \cap S(\tau_2) = \varnothing$ for all $\tau_1 \neq \tau_2$, $\tau_1, \tau_2 \in T$; (v) for all $\tau_2 > \tau_1 \in T$ and any timelike separated pair of points $p_2 \in S(\tau_2)$ and $p_1 \in S(\tau_1)$, p_2 is to the future of p_1.

The idea is that successive spacelike hypersurfaces in an S-partition are ordered (in a relevant way) in the timelike direction.

The objection to MURIC** from the possibility of non–flat IR-gaps could then be met by upgrading MURIC** to MURIC***:

(MURIC***)   If an object *o* is q-located at two spacetime regions, then no matter whether or not the contents of these regions *themselves* are related by immanent causation, it must be the case that these regions are slices through the path *o* of *o*, such that for *any* S-partition $\{S(\tau), \tau \in T\}$ of the

---

[53] See Section §5.2 above for a detailed discussion of this issue. Non-flat spacelike hypersurfaces in Minkowski spacetime have a number of interesting properties, which are further discussed in §7.9 and Appendix 7.1 to Chapter 7.

[54] Obviously, this deviates from (FLAT) (§5.2). This is particularly evident in MURIC***, which extends—implausibly, I maintain—q-locations of persisting objects to non-flat achronal regions. Such temporary deviation is adopted below solely for the sake of discussion.

> relevant section of o, the appropriate sort of immanent causal relation holds between the contents of $S(\tau_1)$ and $S(\tau_2)$ for all $\tau_1 \neq \tau_2$, $\tau_1, \tau_2 \in T$.

Here T is the range of the τ-parameter corresponding to "the relevant section of o," which must include the two regions in question.

As just noted, the "appropriate sort of immanent causal relation," which must hold between the contents of $S(\tau_1)$ and $S(\tau_2)$, cannot, in the case of MURIC***, be associated with the laws of relativistic dynamics (for $S(\tau_1)$ and $S(\tau_2)$ do not, in general, represent any dynamically significant notions in SR). Rather it must be gleaned from those considerations that give force to the objection, namely, to the claim that if the careers of a and a' are separated by a spacelike, but not necessarily flat, macro-IR-gap, then the relevant states of a and a' are not related by immanent causation.

It is easy to see that, just like MURIC**, MURIC*** gives a correct verdict in the case of crisscrossing slices (see Figure 5.7). (I shall not go over it here.) Unlike MURIC**, however, MURIC*** is well equipped to handle the objection based on non-flat IR-gaps. To see this, consider an S-partition of the union of a and a' such that, for some $\tau = \tau_{\text{gap}}$, $S(\tau_{\text{gap}})$ includes all the points at which the life careers of the particles composing a terminate. By reasons outlined just above, no immanent causal relation holds between the contents of $S(\tau_{\text{gap}})$ and $S(\tau)$, for all $\tau > \tau_{\text{gap}}$, and therefore, by MURIC***, no single object can be q-located at any two slices of the union of a and a' that are separated by an IR-gap: a highly desirable result.

I conclude that, when appropriately modified, the plausible principle imposing a natural restriction on multi-q-location of enduring or exduring objects in Minkowski spacetime correctly allows the q-location of a single object at crisscrossing regions and, at the same time, correctly disallows it at regions separated by an IR-gap.

In addition, I wish to make a different comment about scenarios involving immaculate replacement. It would appear that, in the context of an ontological debate firmly situated in the framework of a particular physical theory, it could be dialectically appropriate simply to abstract from physically impossible counterexamples. The immaculate replacement scenarios are in this category. The careers of physical objects represented in them (namely, the objects going out of and popping into existence) violate the conservation laws of physics, whereas the relativistic considerations underlying the whole discussion assume their validity. This is not to say that such cases are impossible *tout court*, only to note that their relevance to the debate about persistence in the context of SR could be questioned on rather general grounds.

5.7. Conclusion

In this chapter we have stated various views of persistence in Minkowski spacetime and examined several considerations, old and new, against the mere possibility or coherence of such statements in the cases of endurance and exdurance. None of these considerations have been found to be successful.

No special arguments have been offered to defend the possibility of perdurance in the special relativistic setting. But none are needed, for the four-dimensional framework of Minkowski spacetime populated with the worldlines and worldworms of persisting objects provides a very friendly habitat for perdurance, in the first place. Indeed, as early discussions show, it is all too easy to reify those worldlines and worldworms and jump to the conclusion that they represent, not merely the careers of persisting objects, but the objects themselves.

Our extensive discussion of the various modes of location and multi-location in spacetime should suffice to discourage anyone from making such a move. None of the major views of persistence is a non-starter in Minkowski spacetime, and none seems to be seriously handicapped. If any view enjoys a privilege, more work is needed to demonstrate it. The following chapters take up this task.

6

Coexistence in Spacetime

6.1. The Notion of Coexistence

Any view of persistence must incorporate an interesting notion (or notions) of coexistence. By "interesting," I mean a notion that is neither empty nor universal. In particular, whether objects endure, exdure, or perdure, there must be a sense of the coexistence relation such that I coexist with Nicolas Sarkozy but not with Napoleon. This much I take to be uncontroversial.

To be sure, if one is an eternalist one must admit that there is a basic sense in which *all* the inhabitants of spacetime—those which we call (in our unreflective moments) past, present, and future—coexist with each other, simply because all such entities—including dinosaurs, Nebuchadnezzar, Napoleon, Nicolas Sarkozy, future human outposts on Mars, and Gorbachev's great great grandson—*exist*, at their respective spacetime locations. But even the eternalist must admit that there is another and non-trivial sense of the coexistence relation such that she bears this relation to Sarkozy but not to Nebuchadnezzar. Sarkozy, in his turn, bears this relation to the eternalist, but there was a time when he did not. Sarkozy never coexisted, in the interesting sense, with Napoleon; Sarkozy no longer coexists with Ronald Reagan, but there was a time when he still did.

This sense of fleeting coexistence is familiar and important and its application is not limited to sentient beings. The Great Wall of China coexists, in this interesting sense, with the Taj Mahal, but there was a time when it did not. The Milky Way has similarly coexisted (and still does—as far as we know) with the Magellanic Clouds.

It is also clear that non-trivial coexistence is a multigrade relation. Things are capable of interesting coexistence, not just pair-wise, but collectively or *en masse*. Sarkozy, Putin, and Obama now stand in a single three-place non-trivial relation of coexistence with each other, and 100,127 select stars in the Milky Way stand in a single such 100,127-place relation.

Does the non-trivial relation of coexistence cut any ontological ice, or is it a merely (or primarily) "locative" notion? The question is important, and I shall return to it in §7.1. For now, let me simply note that, whether or not the interesting sense of coexistence is decidedly ontological or solely locative, it is important. This is easy to see in the classical framework, where the relation of coexistence is especially simple and unproblematic. I discuss it in §6.3. But my real concern is, of course, to extrapolate the interesting notion of coexistence to Minkowski spacetime (§6.4).

To the extent that this is possible it requires, not surprisingly, certain adjustments in the notion itself. (Remember: everyone must be prepared to make adjustments in making a transition to something as full of surprises as Minkowski spacetime.) The need to make the adjustments brings to the surface two distinct strands in the "intuitive lore" of the coexistence relation. They work in unison in the classical setting but come apart in the relativistic context, thus giving rise to two different sets of rules associated with the use of the interesting concept of coexistence. Which of them, if either, can claim the title? I consider both alternatives and defend my preferences in §§6.4–6.7.[1]

I start by introducing a general framework for discussion and by imposing certain requirements on the (interesting) notion of coexistence in spacetime.

6.2. Desiderata

I contend that interesting coexistence[2] of two arbitrary objects must be grounded in an objective relation between their spatiotemporal locations. Such a relation defines the basic notion of coexistence, which, in turn, gives rise to a family of derivative notions, some familiar some new. For a pair of enduring or exduring objects, the basic notion, CE, is a four-place

---

[1] My preferred account of coexistence in Minkowski spacetime was first introduced in Balashov 2000a, b and further developed in Balashov 2002 and 2005. Parts of §§6.2–6.4 build on this previous work. The alternative account is due to Gibson and Pooley 2006.

[2] From now on, I shall frequently drop the qualification 'interesting.'

relation involving two objects o_1 and o_2 and their respective spatiotemporal q-locations p_1 and p_2.[3] CE obtains just in case o_1 and o_2 are q-located at spacetime points p_1 and p_2 respectively and the relevant two-place relation holds between p_1 and p_2:

(CE) $CE(o_1, o_2, p_1, p_2) =_{df} o_1$ is q-located at $p_1 \wedge o_2$ is q-located at p_2
$\wedge \Sigma(p_1, p_2)$.

The nature of Σ is determined by a particular spacetime theory.

What about perduring objects? The ground-level notion of coexistence appropriate for them is the dyadic relation CP between the locations of their momentary diachronic parts:

(CP) $CP(p_{\|1}, p_{\|2}) =_{df} \Sigma(p_1, p_2)$

Putting matters this way helps to state some natural requirements that must be satisfied by the (interesting) notion of coexistence in spacetime.

(Symmetry) If o_1 (as at p_1) coexists with o_2 (as at p_2) then o_2 (as at p_2) coexists with o_1 (as at p_1).[4]

Any notion of coexistence violating Symmetry would be a non-starter and would hardly deserve its title.

(Objectivity) Given two objects and their q-locations in spacetime, there must be a fact of the matter about their coexistence.

In the context of many familiar spacetime theories Objectivity is synonymous with *frame invariance* and opposed to frame relativity. This is not to deny the importance of many frame-relative properties (e.g. velocity, acceleration, kinetic energy), only to stress that coexistence must not be frame-relative. This is also not to deny that there is a perfectly good sense in which frame-relative facts are objective, but to stress that coexistence must be objective in a different sense.

---

[3] Recall (§2.4.2) that an object is (exactly) q-located at a spacetime region iff either the object itself or its non-modal counterpart is (exactly) located there. Both enduring and exduring objects persist by being q-located at multiple 3D regions.

[4] 'o_i (as at p_i)' refers to o_i, as q-located at p_i, in the case of enduring and exduring objects, and to the diachronic part of o_i (i.e., to $p_{\|i}$), in the case of perduring objects. This harmless stretch of terminology will allow us to keep the statements of the Symmetry, Objectivity, and Relevance sufficiently general and neutral among the rival views of persistence.

Objectivity closes some doors.[5] But I think those doors must be closed. As we shall see below, Objectivity keeps other important doors open.

(Relevance): $\Sigma(p_1, p_2)$ must be relevant to the coexistence of o_1 and o_2 (as at p_1 and p_2, respectively).

Relevance needs some clarification. There are many *objective* relations between spatiotemporal locations or q-locations of persisting objects (or their diachronic parts). But most of them express facts that are not in the least relevant to the coexistence of such objects. For example, 'p_1 is two hundred thousand miles closer to the Titanic's sinking point than p_2 and five hundred years later than p_2, as measured in Alpha Centauri's rest frame' is such an objective relation which is completely irrelevant. If the coexistence of o_1 and o_2, or of their diachronic parts, is understood as grounded in some relation $\Sigma(p_1, p_2)$ holding between their (q-)locations p_1 and p_2, it had better be clear why this relation's holding means that o_1 and o_2 coexist, as at p_1 and p_2.[6]

(Multigrade) Coexistence in spacetime must be a multigrade relation or capable of being upgraded to a multigrade relation.

This has already been noted: objects must be capable of interesting coexistence, not just pairwise, but *en masse*, by virtue of standing in a single *n*-grade relation to each other. As stated, CE and CP are two-grade relations. But in familiar contexts they can easily be upgraded to *n*-grade relations for any arbitrary *n*.[7]

(CE_n) $CE_n(o_1, o_2, \ldots o_n, p_1, p_2, \ldots p_n) =_{df} o_1$ is q-located at $p_1 \wedge o_2$ is q-located at $p_2 \wedge \ldots o_n$ is q-located at $p_n \wedge \Sigma(p_1, p_2, \ldots p_n)$.

(CP_n) $CP_n(p_{\|1}, p_{\|2}, \ldots p_{\|n}) =_{df} \Sigma(p_1, p_2, \ldots p_n)$.

---

[5] In particular, it disqualifies certain interesting proposals that are explicitly grounded in frame-relative structures, such as Cody Gilmore's recent suggestion (Gilmore 2002: 254) to relativize coexistence in Minkowski spacetime to a hyperplane of simultaneity (HPS):

(REL) For any two objects in Minkowski spacetime and any HPS, the objects in question coexist *at* this HPS iff their paths (worldlines) intersect it.

For a detailed discussion of REL, see Balashov 2005c.

[6] I am indebted to Gibson and Pooley 2006: 169 for this way of putting Relevance.

[7] The terms 'two-grade' and '*n*-grade,' as they are used here and below, express the sense in which coexistence relates two or *n*>2 objects, which should be distinguished from the more technical sense in which CE_n is, strictly speaking, a $2n$-place relation (and $CE_n{}^*$ is an $n+1$-place relation).

When so upgraded, the resulting relations must, of course, satisfy the n-place counterparts of Symmetry, Objectivity, and Relevance, as follows:

(Symmetry$_n$) If o_1 (as at p_1), o_2 (as at p_2), . . . and o_n (as at p_n) coexist then $o_{i(1)}$ (as at $p_{i(1)}$), $o_{i(2)}$ (as at $p_{i(2)}$), . . . and $o_{i(n)}$ (as at $p_{i(n)}$) coexist for any permutation $\{i(1), i(2), \ldots i(n)\}$ of the set $\{1, 2, \ldots n\}$.

(Objectivity$_n$) Given n objects and their q-locations in spacetime, there must be a fact of the matter about their coexistence.

(Relevance$_n$) $\Sigma(p_1, p_2, \ldots p_n)$ must be relevant to the coexistence of $o_1, o_2, \ldots$ and o_n (as at $p_1, p_2, \ldots$ and p_n, respectively)

I contend that Symmetry, Objectivity, Relevance, and Multigrade are reasonable requirements to impose on any interesting notion of coexistence in spacetime.

Let us see how these ideas work in the classical setting.

6.3. Coexistence in Galilean Spacetime

In Galilean spacetime (ST^G), the coexistence-grounding relation $\Sigma(p_1, p_2)$ is simply that of belonging to the same absolute time hyperplane:

$$\Sigma^G(p_1, p_2) =_{df} t_1 = t_2$$

where, as usual, (x_1, y_1, z_1, t_1) and (x_2, y_2, z_2, t_2) are the coordinates of p_1 and p_2 in any Cartesian coordinate system associated with any inertial frame of reference.

On CE^G, then, two enduring or exduring objects coexist just in case their momentary q-locations are contemporaneous, or co-present:

(CE^G) Two enduring or exduring objects o_1 and o_2, q-located at points p_1 and p_2 of ST^G, respectively, *coexist* $=_{df}$ p_1 and p_2 belong to the same moment of absolute time:

$$CE^G(o_1, o_2, p_1, p_2) =_{df} t_1 = t_2$$

The corresponding perdurantist principle is as follows:

(CP^G) A diachronic part $p_{\|1}$ of a perduring object o_1, located at point p_1 of ST^G, *coexists* with a diachronic part $p_{\|2}$ of a perduring object

o_2, located at point $p_2 =_{df} p_1$ and p_2 belong to the same moment of absolute time:

$$CP^G(p_{\|1}, p_{\|2}) =_{df} t_1=t_2.$$

Spacetime locations and q-locations of objects (or their diachronic parts) in ST^G can be conveniently individuated by absolute dates or, in some cases, by the ages of the objects in question. For example, the 47-year-old enduring president Obama coexists, in the basic sense of CE^G, with the 54-year-old president Sarkozy, but not with the 49-year-old Sarkozy. Alternatively (if less elegantly), the enduring Obama, when he is located at some moment in 2009, coexists with Sarkozy, when the latter is located at the same moment, but not when Sarkozy is located at any moment in 2004.

Admittedly, the basic relation of coexistence CE^G (and, to a lesser extent, CP^G) is a theoretical concept that is at some remove from the more intuitive notions. But it enjoys the sort of generality that allows it to serve as a common ground for various more intuitive notions. For example, a more recognizable triadic relation 'o_1 coexists with o_2 at t' can be defined as '$(\exists p_1, p_2)$ (o_1 is q-located at $p_1 \wedge o_2$ is q-located at $p_2 \wedge t_1=t_2=t$)'. Obama and Sarkozy coexist, in this sense (at any moment) in 1966 but not in 1956. There is, in fact, an entire family of distinct notions of coexistence, some familiar some new (they will emerge shortly), which all have their basis in CE^G.[8]

One important derivative sense of coexistence is that in which an enduring or exduring object o_1, wholly present (i.e., located or q-located) at a certain moment of its proper time t (in the classical case, intervals of proper time coincide with intervals of the common universal time), coexists with another enduring or exduring object o_2 taken in abstraction from any of its q-locations. I shall refer to this derivative sense as *coexistence**, to distinguish it from the more basic (but less familiar) sense of coexistence expressed by CE. Classically, o_1, *when* it is wholly present, or q-located, at a point belonging to some moment of absolute time, coexists* with o_2 just

---

[8] In this respect, CE is similar to the concept of spatial location, which is normally taken to be fundamental; more empirical notions such as distance, angles in space, and so on can then be defined on its basis. Although in the classical setting one could do without CE^G and start, say, with the more familiar triadic relation 'o_1 coexists with o_2 at t,' CE is indispensable in the relativistic framework, in which no common 't' is generally available to a pair of objects.

in case o_2 is wholly present, or q-located, at some point belonging to the same moment:

> (CE^{G*}) An enduring or exduring object o_1 q-located at p_1 *coexists** with o_2 in ST^G $=_{df}$ o_2 is q-located at a point p_2 belonging to the same absolute moment of time as p_1:
>
> CE^{G*}(o_1, p_1, o_2) $=_{df}$ $\exists p_2$ (o_2 is q-located at p_2 $\wedge$ $t_1=t_2$).

I at 1961 (read: when I am wholly present at some moment in 1961) do, but I at 1968 do not coexist* with JFK (taken in abstraction from any of his spatiotemporal q-locations). I at 1961 do not coexist* with FDR. At any moment t of my life, all enduring or exduring objects in the universe, with which I ever coexist, can be divided into three classes: those with which, intuitively speaking, I *no longer* coexist; those with which I *still* or *already* coexist; and those with which I do *not yet* coexist. As I grow older, the membership of these classes changes, thanks to the presence in our world of many enduring or exduring objects that come to be and cease to exist. In all cases, the relation underwriting these temporally laden determinations (i.e., *not yet*, *still*, *already*, and *no longer*) is coexistence*.

In addition to coexistence*, there is also a familiar temporally unmodified sense in which I (also considered in abstraction from any of my momentary q-locations) never coexist with FDR but do coexist (in the same sense) with JFK. This sense—coexistence** or "temporal overlap"—is arrived at by a further generalization of classical coexistence*:

> (CE^{G**}) Two enduring or exduring objects o_1 and o_2 *coexist** ("temporally overlap") in ST^G $=_{df}$ o_1 and o_2 have q-locations belonging to the same moment of absolute time:
>
> CE^{G**}(o_1, o_2) $=_{df}$ $\exists p_1$, p_2 (o_1 is q-located at p_1 $\wedge$ o_2 is q-located at p_2 $\wedge$ $t_1=t_2$).

The perdurantist analogs of coexistence* and coexistence** are readily available.

> (CP^{G*}) A diachronic part $p_{\|1}$ of a perduring object o_1, located at point p_1, *coexists** with another perduring object o_2 in ST^G $=_{df}$ o_2 has a diachronic part $p_{\|2}$ located at p_2, such that $t_1=t_2$.
>
> CP^{G*}($p_{\|1}$, o_2) $=_{df}$ $\exists p_{\|2}$ ($p_{\|2}$ is located at p_2 $\wedge$ $t_1=t_2$).

(CP^{G**}) Two perduring objects o_1 and o_2 *coexist*** ("temporally overlap")
in ST^G $=_{df}$ o_1 and o_2 have diachronic parts $p_{\|1}$ and $p_{\|2}$,
respectively, that are located at the same moment of absolute
time:

CP$^{G**}=_{df}$ $\exists p_{\|1}, p_{\|2}$ ($p_{\|1}$ is located at p_1 $\wedge$ $p_{\|2}$ is located at
p_2 $\wedge$ $t_1=t_2$).

In ordinary situations 'coexistence' is often used to express the senses of
coexistence* and coexistence** and, less often, the sense of coexistence or
some notion in its neighborhood. Below I will allow myself the liberty to
use these different varieties of the interesting notion of coexistence without
labeling their notations with asterisks in cases where context makes it clear
which notion is at work.

Coexistence in ST^G, as defined by Σ^G, is symmetrical, objective, and
relevant. It is also multigrade-ready; that is to say, it can easily be extended
to the case of $n>2$ objects bearing to each other a *single* relation of
coexistence. For n objects, the coexistence-grounding relation $\Sigma_n{}^G$ is still
that of belonging to the same absolute time hyperplane, but now holding
among n spacetime points:

$$\Sigma^G(p_1, p_2, \ldots p_n) =_{df} t_1=t_2=\ldots=t_n.$$

Stating the resulting coexistence principles[9] is tedious and is relegated to
Appendix 6.1 at the end of this chapter.

When so upgraded, classical coexistence relations satisfy the n-grade
analogues of Symmetry (where appropriate), Objectivity, and Relevance.
Relevance, in particular, is observed because coexistence in classical spacetime has a clear intuitive link to the notion of *co-presence* at a single moment
of absolute time.

This notion, however, is not available in Minkowski spacetime. What
could play its role?

6.4. Coexistence in Minkowski Spacetime: CASH

I submit that the best candidate to ground coexistence in Minkowski
spacetime (ST^M) is the invariant relation of *belonging to a common time*

---

[9] That is, (CE$_n{}^G$), (CE$_n{}^{G*}$), (CE$_n{}^{G**}$), (CP$_n{}^G$), (CP$_n{}^{G*}$), and (CP$_n{}^{G**}$).

hyperplane. I shall refer to this relation as CASH (for Coexistence As Sharing a Hyperplane of simultaneity):

$$\Sigma^M{}_{CASH}(p_1, p_2) =_{df} \exists F \; t^F{}_1 = t^F{}_2$$

Here F is an inertial reference frame and $t^F{}_1$, $t^F{}_2$ moments of time in a Cartesian coordinate system adapted to F.[10]

CASH emerges as the best candidate for a number of reasons (see Balashov 2000a: 132–52). Here I wish to note that, once we settle on Objectivity, the range of choices becomes rather limited, for the relevant relation must be based on the invariant structures of Minkowski spacetime. Spacelike separation then appears to be the most plausible option, which also correctly recovers the classical limit (i.e. the notion of co-presence at a moment of absolute time), of which more below.

CASH gives rise to the following relativistic analogues of the classical accounts of coexistence.

($CE^M{}_{CASH}$) Two enduring or exduring objects o_1 and o_2, q-located at points p_1 and p_2 of ST^M respectively, *coexist* $=_{df}$ p_1 and p_2 belong to a common time hyperplane:

$CE^M{}_{CASH}(o_1, o_2, p_1, p_2) =_{df} \Sigma^M{}_{CASH}(p_1, p_2)$.

($CE^{M*}{}_{CASH}$) Enduring or exduring object o_1 q-located at spacetime point p_1 *coexists** with o_2 considered in abstraction from its q-location in ST^M $=_{df}$ o_2 is q-located at a spacetime point that shares a common time hyperplane with p_1:

$CE^{M*}{}_{CASH}(o_1, p_1, o_2) =_{df} \exists p_2$ (o_2 is q-located at $p_2 \wedge \Sigma^M{}_{CASH}(p_1, p_2)$).

($CE^{M**}{}_{CASH}$) Two enduring or exduring objects o_1 and o_2 *coexist*** ("temporally overlap") in ST^M $=_{df}$ o_1's and o_2's paths have points sharing a common time hyperplane:

[10] A predecessor of CASH was introduced in Balashov 2000a, 2000b and later referred to by Gilmore (2002) as CASS (Coexistence As Spacelike Separation). Gilmore rightly noted that CASS was deficient (and, therefore, 'CASS' was a misnomer) in that spacelike separation is an anti-reflexive relation that rules out a trivial case of coexistence between o located at p and itself. The relation that does the job right is the disjunctive relation *spacelike-separation-or-spacetime-coincidence*, $I(p_1, p_2) < 0 \vee p_1 = p_2$ (where $I(p_1, p_2) \equiv c^2(t_2 - t_1)^2 - (x_2 - x_1)^2$ is the relativistic interval), which is equivalent to $\Sigma^M{}_{CASH}(p_1, p_2)$, as defined above. $\Sigma^M{}_{CASH}(p_1, p_2)$, however, is more elegant and can be naturally generalized to an n-grade relation (see below), so we shall use it henceforth. For a more detailed discussion of CASS and CASH, see Balashov 2005c.

$\mathrm{CE^{M**}_{CASH}}(o_1, o_2) =_{df} \exists p_1, p_2 \ (o_1 \text{ is q-located at } p_1 \wedge o_2 \text{ is q-located at } p_2 \wedge \Sigma^M_{CASH}(p_1, p_2)).$

$(\mathrm{CP^M_{CASH}})$ Diachronic parts $p_{\|1}$ and $p_{\|2}$ of perduring objects o_1 and o_2, respectively, *coexist* in $\mathrm{ST^M} =_{df}$ The locations of $p_{\|1}$ and $p_{\|2}$ belong to a common time hyperplane:

$\mathrm{CP^M_{CASH}}(p_{\|1}, p_{\|2}) =_{df} \Sigma^M_{CASH}(p_1, p_2).$

$(\mathrm{CP^{M*}_{CASH}})$ A diachronic part $p_{\|1}$ of a perduring object o_1, located at point p_1, *coexists** with another perduring object o_2 in $\mathrm{ST^M}$ $=_{df}$ o_2 has a diachronic part $p_{\|2}$ located at point p_2, which shares a common time hyperplane with p_1:

$\mathrm{CP^{M*}_{CASH}}(p_{\|1}, o_2) =_{df} \exists p_{\|2} \ (p_{\|2} \text{ is located at } p_2 \wedge \Sigma^M_{CASH}(p_1, p_2)).$

$(\mathrm{CP^{M**}_{CASH}})$ Two perduring objects o_1 and o_2 *coexist*** ("temporally overlap") in $\mathrm{ST^M} =_{df}$ o_1 and o_2 have diachronic parts $p_{\|1}$ and $p_{\|2}$, respectively, that share a common time hyperplane:

$\mathrm{CP^{M**}_{CASH}}(o_1, o_2) =_{df} \exists p_{\|1}, p_{\|2} \ (p_{\|1} \text{ is located at } p_1 \wedge p_{\|2} \text{ is located at } p_2 \wedge \Sigma^M_{CASH}(p_1, p_2)).$

There is a great deal to say in favor of CASH-based relativistic coexistence. It is obviously objective and symmetric.[11] On CASH, lots of objects in the universe interestingly coexist and lots of objects interestingly fail to coexist. And in most cases CASH-based verdicts are in perfect agreement with intuitive verdicts. This underscores the relevance of CASH. Indeed, the relevance has to do with CASH's correct *classical limit*. This is only to be expected, given that the classical limit of $\Sigma^M_{CASH}(p_1, p_2)$, the relation of sharing a common time hyperplane, is just $\Sigma^G(p_1, p_2)$, the relation of belonging to the same moment of absolute time.

Last but not least, CASH is easily upgradeable to an *n*-grade relation CASH_n based on an *n*-adic extension of $\Sigma^M_{CASH}(p_1, p_2)$:

$$\Sigma^M_{CASH}(p_1, p_2, \dots p_n) =_{df} \exists F \ t^F_1 = t^F_2 \dots = t^F_n$$

---

[11] Where symmetry is applicable, i.e. in the cases of $\mathrm{CE^M_{CASH}}$, $\mathrm{CE^{M**}_{CASH}}$, $\mathrm{CP^M_{CASH}}$, and $\mathrm{CP^{M**}_{CASH}}$, but not in the cases of $\mathrm{CE^{M*}_{CASH}}$ and $\mathrm{CP^{M*}_{CASH}}$.

Again, F is an inertial reference frame and t^F_i a moment of time in a Cartesian coordinate system adapted to F. The resulting coexistence principles are stated in Appendix 6.2.

There is a great deal to say in favor of the n-grade CASH-based relativistic coexistence. It is obviously symmetric (in an appropriate n-adic sense) and objective. It is also relevant. On $CASH_n$, many objects in the universe interestingly coexist (coexist*, coexist**), not just pairwise, but *en masse* (as they should, intuitively speaking) and many objects interestingly fail to coexist (in a similar way). For example, all the pieces of furniture in my office, all the people currently alive, all the planets in the Solar system, and all the *Star Trek* characters can bear a *single* $CASH_n$-based relation of coexistence to each other. The same is true of the trio of astronomers, Galileo, Kepler, and Brahe. Adding Ptolemy to the group, however, changes the situation rather dramatically. In general, all groups of objects that coexist (coexist*, coexist**) or fail to coexist (coexist*, coexist**) classically continue to do so relativistically, which shows that CASH has a correct classical limit. Again, this is anything but surprising.

To be sure, there are some notable differences between the classical and relativistic coexistence. For example, if n objects stand in a single n-grade relation of coexistence (governed by $CASH_n$) in Minkowski spacetime then, taken pairwise, they also stand in $n(n-1)/2$ two-grade relations of coexistence (governed by CASH). This is only to be expected, classically speaking. But the converse does not hold.[12] While this may be a bit surprising it should not be overdramatized. If anything, it shows that the total CASH value of collective coexistence in ST^M is greater than the sum of its piecemeal two-place values.

Where CASH and $CASH_n$ seem to differ most from their classical counterparts is in regard to the issue of transitivity. Classically, if o_1 (as at p_1) coexists with o_2 (as at p_2) and the latter coexists with o_3 (as at p_3) then o_1 (as at p_1) coexists with o_3 (as at p_3). This result does not transfer to ST^M. Some will, no doubt, say that this amounts to a total fiasco:

---

[12] This has to do with the fact that for $n > 4$, if spacetime points $p_1, p_2, \ldots p_n$ in ST^M are pairwise spacelike separated, it does *not* follow that $p_1, p_2, \ldots p_n$ belong to a common time hyperplane. Compare this to a similar geometrical fact about $n > 3$ points in the usual three-dimensional space.

what is coexistence without transitivity? But I am not inclined to be so pessimistic.

First, while transitivity may appear to be an indispensable intuitive feature of the interesting notion of coexistence, this is only when the latter is considered in a wholesale and unspecified way. For one thing, transitivity is a property of dyadic relations, while the core *intuitive* notion of coexistence is *triadic*: 'o_1 coexists with o_2 at t.' Granted, there is a sense in which a triadic relation can be viewed as being transitive when one of its terms is held fixed, and perhaps that is precisely the sense in which *one* of the intuitive classical notions of coexistence is transitive. On the other hand, there are contexts in which other notions of coexistence come to the forefront and force transitivity to break down *even* in the classical case. I coexist (in the sense of coexistence** or "temporal overlap") with JFK and he coexists (in the same sense) with FDR. But I do not coexist with FDR. This holds true both classically and relativistically, for all types of objects, enduring, exduring, or perduring. Thus we are familiar with an important sense in which the temporally *un*modified relation of coexistence is not transitive. CASH then extends this feature to the temporally modified four-place relativistic relation of coexistence governed by CE^M_{CASH} (and to the two-place relation governed by CP^M_{CASH}). In view of the initial plausibility of CASH, accepting this increased degree of non-transitivity (over against the classical case) is, I submit, sound strategy. After all, every account has some costs. Note also that if the basic relations governed by CE^M_{CASH} and CP^M_{CASH} were, *per impossibile*, transitive, they would be universal relations. And this is surely to be resisted. I today do not coexist with myself tomorrow. Neither do I coexist with the big bang. *Mutatis mutandis* for my diachronic parts.

But second, any residual worry one might have about the absence of transitivity at the level of the basic relation defined by CASH evaporates when CASH is upgraded to CASH$_n$. The problem of transitivity arises only when the number of objects under consideration exceeds two; but CASH$_n$ is better equipped (than CASH) to handle all such cases (where $n > 2$) to begin with, by offering a *single* n-grade relation of coexistence for them to stand in.

I conclude that, overall, CASH does a good job grounding an interesting—symmetrical, objective, and relevant—relation of coexistence in Minkowski spacetime. There is a clear sense in which it extrapolates the corresponding classical relation to a new spacetime framework. There is a clear sense in which CASH expresses the idea of co-presence, without invoking the notion of *the* present. And there is a clear sense in which CASH has a correct classical limit. With these credentials, CASH is, arguably, the best candidate for the job. But this claim has not gone unchallenged. CASH has a strong competitor.

6.5. Alexandrov-Stein Present and Alexandrov-Stein Coexistence

The competitor takes its inspiration from an ingenious way of analyzing tensed utterances in the eternalist spacetime framework. Any such analysis must delineate the *scope of the present*, relative to a token utterance or thought. In the classical framework, it is natural to associate the scope of the present with the set of events that are simultaneous with the utterance (or thought). Depending on the details of the analysis, this set may include just a single time hyperplane in Galilean spacetime or a "stack" of such hyperplanes. The second option is favored by the friends of the "specious present"—those who think that the extent of the present is a short but finite interval of time associated, one way or the other, with the perception of "now."

The next question is how to extrapolate this idea to Minkowski spacetime. One strategy that might look initially attractive is to identify the present with a time hyperplane (or a family of such hyperplanes) comprising the events that are simultaneous with the utterance (or thought) in the instantaneous rest frame of the subject.[13] This approach, however, confronts a famous problem[14] that two speakers in relative motion will disagree about the scope of the distant present. For one such speaker (located on a remote star) World War II may *already* be over (as it is for all of us), while for the other it may *still* be going on. To avoid such a problem, the scope of the present ought not to be relativized to the state of motion of the speaker; it must be frame-invariant.

Another strategy—one used in CASH—is, in effect, to grant the status of the present indiscriminately to all the relevant time hyperplanes—those

---

[13] For a sketch of such a proposal, albeit in the context of stage theory of persistence, see Sider 2001: 198–9.

[14] Variously exploited in the "Putnam–Stein debate."

containing the momentary location of the speaker (or any other object, for that matter), and then identify one of them as constituting a common present for a collection of objects.

But is it necessary to associate the present with a spatially unbounded time hyperplane (or a family of such) in Minkowski spacetime? Must the scope of the present be infinite in extent? Some authors believe this is not only unnecessary but, in fact, inappropriate. On the view adopted and developed by Gibson and Pooley (2006: 166 ff), the present comprises the locations[15] of the entities with which the speaker can *interact* (i.e. affect and be affected by) during a short but finite interval of proper time Δt^{NOW} required to have a single thought or experience (perhaps 0.2 sec). Objects that can affect the speaker during such "specious present" are located in the past lightcone of the upper boundary of Δt^{NOW}, while objects that can be affected by her are located in the future lightcone of the lower boundary of Δt^{NOW}. Objects that can do *both* are therefore located at the intersection of these two regions. Let us refer to this intersection as $\text{Present}_{\text{AS}}(\Delta t^{\text{NOW}})$ —the *Alexandrov-Stein Present* of Δt^{NOW}.[16]

The Alexandrov-Stein Present (AS-Present) of Δt^{NOW} is a discus-shaped region with a huge spatial extent (about 60,000 km) and a very short temporal extent (0.2 sec), which, in all ordinary contexts, is indistinguishable from an infinite hyperplane (see Figure 6.1). According to the advocates of AS-Present, therein lies the source of common false beliefs about the extent of distant present. The underlying idea is that we always tend to

[15] Or q-locations. The entities in question may be ordinary objects or their diachronic parts, depending on whether objects endure, exdure, or perdure. In the remainder of this chapter I often push these important distinctions to the background. The reason is that the issue under discussion concerns the very basic nature of the (interesting) coexistence relation in Minkowski spacetime (i.e. CASH versus AS), which all the rival theories of persistence must adopt as a starting point. To avoid unnecessary complication it is therefore appropriate to resort, in the present context, to an informal and sufficiently neutral vocabulary of notions such as 'location,' 'occupation,' 'being present at,' and the like.

[16] The 'Stein' part of the terms 'Alexandrov-Stein Present' ('AS-Present') and 'Alexandrov-Stein Coexistence' ('AS-Coexistence'), which occur in this section heading and the ensuing discussion, reflects the importance and influence of related ideas developed by Howard Stein 1991: 158 ff. See, in this connection, Gibson and Pooley 2006: 167, who use the term 'Stein Present.' The 'Alexandrov' part alludes to a different context, in which discus-shaped double lightcone regions of the type described above are known as "Alexandrov (Alexandroff) neighborhoods" or "Alexandrov (Alexandroff) intervals." They figure, in particular, in classical sources on the foundations of general relativity, in contemporary literature on quantum gravity, and in Rafael Sorkin's "causal sets program"; see, e.g. Sorkin 1991. They were also used by John Winnie in his influential derivation of the geometrical structure of Minkowski spacetime from its causal structure (Winnie 1977: 156 ff). See also Arthur 2006 and Savitt 2009.

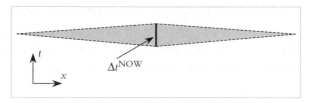

Figure 6.1. Alexandrov-Stein Present of Δt^{NOW}.

associate such extent with the range of entities with which we can interact. In the classical context (which assumes both the notion of absolute time and the idea of arbitrarily fast causal signals), this range is indeed an infinite time hyperplane (or a "stack" of such). In the relativistic context, the range of the present is, strictly speaking, the AS-Present of the specious NOW, which, for the reasons having to do with its large spatial extent and short temporal extent, is easily mistaken for an infinite time hyperplane. There is then a clear sense in which Alexandrov-Stein Present has a correct "classical limit." This allows AS-Present to serve as a basis for an error theory of ordinary beliefs about the extent of the present.[17]

The concept of AS-Present also provides a ground for a non-trivial alternative to CASH. Suppose two objects o_1 and o_2 are located at p_1 and p_2, respectively. Assuming that one can associate with o_1 and o_2 the corresponding finite intervals of their extended NOWs, Δt^{NOW}_1 and Δt^{NOW}_2, let us say that they *Alexandrov-Stein-coexist* (*AS-coexist*) iff their locations fall within each other's AS-Presents (see Figure 6.2).

The basic idea is that objects coexist (in an interesting sense) provided their presents "substantially overlap" and, hence, "there is a reasonable

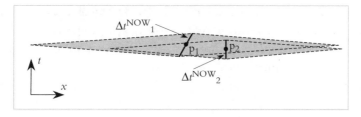

Figure 6.2. Alexandrov-Stein Co-presence and Alexandrov-Stein Coexistence.

---

[17] For important earlier discussions of such an error theory see Butterfield 1984 and Stein 1991: 158 ff. For recent developments, see Arthur 2006; Gibson and Pooley 2006: 166–7, 170–1; and Savitt 2009.

sense in which they share a common present" (Gibson and Pooley 2006: 170). To put the idea more precisely, let us introduce the relation of *Alexandrov-Stein Co-presence* $\Sigma^M{}_{AS}(p_1, p_2)$ between the locations (or q-locations) p_1 and p_2 of two objects o_1 and o_2 (of their diachronic parts) that are equipped, at those (q-)locations, with their respective AS-Presents:

$$\Sigma^M{}_{AS}(p_1, p_2) =_{df} p_1 \in Present_{AS}(\Delta t^{NOW}{}_2) \wedge p_2 \in Present_{AS}(\Delta t^{NOW}{}_1).$$

Alexandrov-Stein Coexistence ($CE^M{}_{AS}$, $CP^M{}_{AS}$), Alexandrov-Stein Coexistence* ($CE^{M*}{}_{AS}$, $CP^{M*}{}_{AS}$) and Alexandrov-Stein Coexistence** ($CE^{M**}{}_{AS}$, $CP^{M**}{}_{AS}$) can then be defined as follows:

($CE^M{}_{AS}$) Enduring or exduring objects o_1 and o_2, q-located at spacetime points p_1 and p_2 of ST^M, respectively, *coexist* $=_{df}$ p_1 and p_2 are AS-co-present.[18]

($CE^{M*}{}_{AS}$) Enduring or exduring object o_1 q-located at spacetime point p_1 *coexists* * with o_2 considered in abstraction from its q-locations in ST^M $=_{df}$ o_2 is q-located at a spacetime point p_2 that is AS-co-present with p_1.[19]

($CE^{M**}{}_{AS}$) Two enduring or exduring objects o_1 and o_2 *coexist* ** ("temporally overlap") in ST^M $=_{df}$ o_1's and o_2's paths have AS-co-present points.[20]

($CP^M{}_{AS}$) Diachronic parts $p_{\|1}$ and $p_{\|2}$ of perduring objects o_1 and o_2, respectively, *coexist* in ST^M $=_{df}$ The locations of $p_{\|1}$ and $p_{\|2}$ are AS-co-present.[21]

($CP^{M*}{}_{AS}$) A diachronic part $p_{\|1}$ of a perduring object o_1, located at point p_1, *coexists* * with another perduring object o_2 in ST^M $=_{df}$ o_2 has a diachronic part whose location is AS-co-present with p_1.[22]

[18] Or, more formally: $CE^M{}_{AS}(o_1, o_2, p_1, p_2) =_{df} \Sigma^M{}_{AS}(p_1, p_2)$.

[19] $CE^{M*}{}_{AS}(o_1, p_1, o_2) =_{df} \exists p_2 (o_2 \text{ is q-located at } p_2 \wedge \Sigma^M{}_{AS}(p_1, p_2))$.

[20] $CE^{M**}{}_{AS}(o_1, o_2) =_{df} \exists p_1, p_2 (o_1 \text{ is q-located at } p_1 \wedge o_2 \text{ is q-located at } p_2 \wedge \Sigma^M{}_{AS}(p_1, p_2))$.

[21] $CP^M{}_{AS}(p_{\|1}, p_{\|2}) =_{df} \Sigma^M{}_{AS}(p_1, p_2)$.

[22] $CP^{M*}{}_{AS}(p_{\|1}, o_2) =_{df} \exists p_{\|2} (p_{\|2} \text{ is located at } p_2 \wedge \Sigma^M{}_{AS}(p_1, p_2))$.

(CP$^{\text{M**}}_{\text{AS}}$) Two perduring objects o_1 and o_2 *coexist*** ("temporally overlap") in ST^{M} =$_{\text{df}}$ o_1 and o_2 have diachronic parts located at AS-co-present points.[23]

Initially Alexandrov-Stein Coexistence (AS-Coexistence) may have much to recommend it, especially for those who are attracted to the notion of the "specious present," for it is this notion that enables Gibson and Pooley to draw a connection between the non-trivial sense of coexistence in Minkowski spacetime and the idea of the mutual location of objects in each other's presents. (It may be recalled that CASH was motivated by a similar idea that an interesting sense of coexistence in spacetime has to do with the notion of co-presence.) In my view, however, AS-Coexistence has many defects and must, in the end, be rejected in favor of CASH. One way to show it is to see how AS-Coexistence fares vis-à-vis the above-noted four desiderata: Symmetry, Objectivity, Multigrade, and Relevance. In the next section I argue that while AS-Coexistence may earn a satisfactory mark for Symmetry, it barely passes Multigrade and seriously stumbles on Objectivity. In §6.7 I show that AS-Coexistence fails the Relevance test on more than one count. The whole situation throws interesting light on the issue of extending important ordinary concepts beyond their original domains.

6.6. AS-Coexistence v. CASH: Symmetry, Multigrade, and Objectivity

The relation of AS-Coexistence is symmetric, even though the more basic relation, associated separately with 'p_1 falls within the Alexandrov-Stein Present of Δt^{NOW}_2' and 'p_2 falls within the Alexandrov-Stein Present of Δt^{NOW}_1', is not—the point noted by Gibson and Pooley (2006: 170). They argue, however, that this is as it should be. Although in many cases the more basic relation holds in both directions, it need not. For two creatures with very different metabolic rates (say, elephants and gnats) and,

[23] CP$^{\text{M**}}_{\text{AS}}(o_1, o_2)$ =$_{\text{df}}$ $\exists p_{\|1}, p_{\|2}$ ($p_{\|1}$ is located at $p_1 \wedge p_{\|2}$ is located at $p_2 \wedge \Sigma^{\text{M}}_{\text{AS}}(p_1, p_2)$).

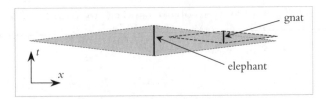

Figure 6.3. The Alexandrov-Stein Presents of a (cosmic) elephant and a (cosmic) gnat.

therefore, different extents of their NOWs (determined, perhaps, by their "specious presents") the basic relation may hold in one directions only (as in Figure 6.3) *precisely* because the "spatial horizons" of the worlds of such creatures, resulting from their respective temporally extended NOWs, are so different.

Yet constructing a symmetrical relation by simply conjoining two non-symmetrical and heterogeneous components[24] leaves one somewhat dissatisfied, and this feeling becomes stronger when AS-Coexistence and its derivatives are upgraded to apply to $n>2$ objects. The sought-for n-grade relation must be constructed as a conjunction of $n(n-1)$ distinct and non-symmetrical relations, as follows:

$$\Sigma^M{}_{AS}(p_1, p_2, \ldots p_n) =_{df} \wedge^n_{i,j=1} \Sigma^M{}_{AS}(p_i, p_j).^{[25]}$$

The detailed structure of the corresponding n-grade relations is spelled out in Appenxix 6.3.

When regimented in this way, the resulting conjunctive n-grade relations $CE_n{}^M{}_{AS}$, $CE_n{}^{M**}{}_{AS}$, $CP_n{}^M{}_{AS}$ and $CP_n{}^{M**}{}_{AS}$ are symmetrical, even though their building blocks are, as before, asymmetrical and heterogeneous. To what extent does this result express the relativistic analogue of the idea of "sharing a common present"—the very idea that, according to Gibson and Pooley, motivates AS-Coexistence in the first place (2006: 170)? I submit that the connection is, at best, very tenuous. There is simply *no* "common present" to share. Instead, there are n distinct presents standing in rather complex relations to each other. Contrast the situation with CASH, where the idea finds a very clear manifestation: a common time hyperplane constitutes a *single* present for an arbitrarily large collection of objects.

---

[24] Heterogeneous—in the sense that each of them involves two different *sorts* of relata, a spacetime *point* and a finite *interval* of proper time (namely, 'p' and 'Δt').

[25] p_i is trivially AS-present with itself for all $i=1, 2, \ldots n$.

Let us now turn to Objectivity. Insofar as the notion of AS-Coexistence is based on the "Alexandrov intervals" of $\Delta t^{NOW}{}_1, \ldots \Delta t^{NOW}{}_n$, it can claim the support of invariant structures of Minkowski spacetime. But by themselves, the structures fall short of fixing the facts about coexistence. Given a collection of objects and their (q-)locations in spacetime, the facts of their coexistence are not determined until $\Delta t^{NOW}{}_1, \ldots \Delta t^{NOW}{}_n$ are supplied, and these do not come from the structure of spacetime itself, but must be added "by hand." Is this acceptable? I return to these questions at the end of the next section.

For now, I want to focus on a different set of problems of a more technical nature. They have to do with attributing AS-Coexistence to entities that lack perception of an extended NOW: rocks, planets, stars, and so forth. It appears obvious that such entities can still coexist with each other in an interesting sense. Mars, for example, coexists, in this sense, with Jupiter most of the time in its life career. But it is unclear how this seemingly objective fact can be expressed in terms of AS-Coexistence. Being aware of this problem, Gibson and Pooley briefly consider what appears to be the next best option: replacing intrinsic facts about coexistence of non-sentient entities in spacetime with the *attribution* of such facts to them by an outside sentient observer. The idea is to associate the extended NOWs of the objects in question with the contextually determined extended NOW of the speaker (ibid.: 170).

But how should such association proceed? Gibson and Pooley do not elaborate, but it appears that one has a choice between two different strategies. I examine them below and show that they both fail, for somewhat different reasons. Consider two objects of the same physical kind, o_1 and o_2, in relative motion and a speaker o (see Figure 6.4) who is in possession of a

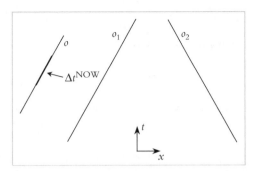

Figure 6.4. Do o_1 and o_2 AS-coexist, as judged by o?

contextually determined interval of the extended NOW, Δt^{NOW}. How is she to attribute Δt^{NOW} to o_1 and o_2? It seems that she can either do it (i) in her own rest frame or (ii) in the rest frames of both objects (in which case Δt^{NOW} will become a common measure of two intervals of proper time of the objects in question). Any other method would be arbitrary.

But the first method will, in some cases, result in a wrong verdict. Consider Figure 6.5, in which the situation looks symmetrical as between o_1 and o_2 located, respectively, at p_1 and p_2. (The situation may represent, say, two perfectly similar stars on course to a head-on collision.) Given that they are objects of the same physical sort, one should expect them to be completely on a par, as far as their coexistence is concerned. But on strategy (i), they are not: while o_2 (as at p_2) falls within the AS-Present of Δt^{NOW}_1, which is equal to Δt^{NOW} (suppose, for simplicity that the speaker and the first object are relatively at rest), o_1 (as at p_1) does *not* fall within the AS-Present of $\Delta t^{NOW}_2 = \gamma\, \Delta t^{NOW}$ (where $\gamma \equiv (1 - v^2/c^2)^{-1/2}$ is the usual relativistic factor). Indeed, p_1 lies outside the Alexandrov-Stein "discus" erected on Δt^{NOW}_2. As a consequence, o_1 and o_2 do not AS-coexist, as at p_1 and p_2. This is hardly acceptable—not because o_1 and o_2 cannot, in general, fail to coexist, but because they should not fail to coexist in such a non-symmetrical fashion (given that they are objects of exactly the same physical sort and are located equally close, in terms of their proper times, to the projected collision point).

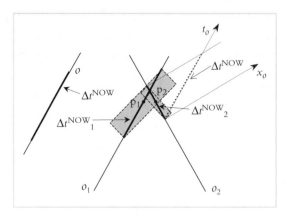

Figure 6.5. p_2 falls within the AS-Present of Δt^{NOW}_1 (assumed equal to Δt^{NOW}), but p_1 does *not* fall within the AS-Present of Δt^{NOW}_2. As a result, o_1 and o_2 do not AS-coexist, as at p_1 and p_2.

COEXISTENCE IN SPACETIME   151

This defect does not upset strategy (ii), which attributes to o_1 and o_2 the same intervals of the extended NOWs in their *own* rest frames: $\Delta t^{NOW}_1 = \Delta t^{NOW}_2 = \Delta t^{NOW}$. But it is unclear what could motivate such attribution. Δt^{NOW} is the interval of the extended NOW of the speaker, which is contextually determined (in some way or other) in *her* rest frame. Simply transferring this determination to the rest frames of o_1 and o_2 does not appear to make any contextual sense. What does the duration of the extended NOW of the speaker in her rest frame have to do with facts about mutual coexistence of insentient objects, when such facts must now be properly grounded in *their* rest frames?

Perhaps the problem could be fixed in some way. But more work needs be done to show it. And the troubles for AS-Coexistence only begin here. More serious problems have to do with the relevance of the underlying relation of AS-co-presence.

6.7. AS–Coexistence v. CASH: Relevance

The chief motivation for AS-Coexistence stems from the idea that co-present objects (i.e. objects located in each other's AS-Presents) must be capable of *interacting* (cf. Gibson and Pooley 2006: 166), and the relation of AS-co-presence is expected to express this idea. One can take issue with both parts of this claim. First, one can argue that the interesting sense of coexistence (and co-presence) has little, if anything, to do with causal interaction. Second, one can argue that the relation of AS-co-presence fails to express the idea of causal interaction between coexisting objects. I shall begin with the second point.

Interaction, in a relevant sense, implies *exchange of energy, information, and so forth. But the complex fact represented by AS-co-presence hardly amounts to that. Indeed, AS-co-presence $\Sigma^M_{AS}(p_1, p_2)$ involves two non-symmetrical relations, $p_1 \in Present_{AS}(\Delta t^{NOW}_2)$ and $p_2 \in Present_{AS}(\Delta t^{NOW}_1)$, which can hold in the *absence* of any communication between the *relevant items*, that is, between o_1, located at p_1, and o_2, located at p_2.[26] If anything, AS-co-presence *precludes* causal communication between *those* items. The

---

[26] Assuming, for simplicity, that objects endure; see n. 15. If objects exdure, replace 'located' with 'q-located.' If objects perdure, replace 'o_1 and o_2' with 'diachronic parts of o_1 and o_2.'

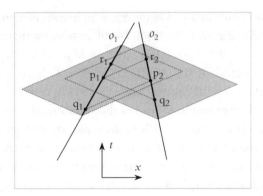

Figure 6.6. Does AS-co-presence really amount to a causal interaction between the relevant items?

way AS-Coexistence handles this situation is by grounding the interesting relation between the items in question (i.e., o_1, located at p_1, and o_2, located at p_2) in the more basic relations involving *additional* items. Object o_2, located at p_2, can receive energy and information from o_1, *when* the latter is located, *not* at p_1, but at an earlier point q_1 in its career. o_2 can then, in its turn, send energy and information back to o_1, which will reach the latter *when* it is located, *not* at p_1, but at a later point r_1 in its career (see Figure 6.6). The same, of course, applies to o_1 and o_2 taken in reverse order (simply replace all occurrences of '1' with '2' and '2' with '1' in the previous two sentences). How does this complicated fact, involving *six* different locations, amount to *interaction* (*communi*cation, *ex*change of energy or information) between o_1 and o_2, *as at p_1 and p_2*? The connection is very remote, at best.

What the complicated fact does show is that, in many cases, objects that coexist in the interesting sense, *when* they occupy their respective locations in spacetime, can *also* causally interact (and exchange energy, information, and so forth many times around) *throughout* more extended periods of time. This suggests that, while both notions—causal interaction and interesting coexistence in spacetime—are important, the relevant link between them may be just the opposite of what AS-Coexistence asserts (and in line with what CASH asserts): objects may coexist *precisely when* they do *not* causally *interact*. This appears to undermine the central idea behind AS-Coexistence (and hence, behind AS-Present).

But is the idea so promising in the first place? "It is intuitively right that, if the temporal extent of the now is of the order of a specious present, then one cannot correctly use the present tense to talk about some event in one's absolute elsewhere that is, say, four light years away," note Gibson and Pooley (2006: 167). Consequently, one cannot correctly speak of the coexistence of objects that are located outside each other's AS-Presents. I simply disagree. In fact, I am inclined to reject both the above conditional and its antecedent. First, there is no need in the extended NOW in anything other than a merely psychological sense. The notion of the extended NOW lacks objective significance.[27] Second, even if the notion of the extended NOW is granted, there is no good reason to refrain from present-tense attributions and to deny coexistence to objects located outside the AS-Present of the speaker (or any other object, for that matter).[28] I shall return to the first point in §6.9. Let me elaborate on the second point.

On Gibson and Pooley's view, the extent of the present (i.e. AS-Present) and, therefore, the range of coexistence (i.e., AS-Coexistence) are spatially limited because these notions have their roots in the idea of the causal interaction between the respective entities. (For the sake of argument we abstract from the above-noted difficulties with the relevance of AS-co-presence to this idea.) CASH, on the other hand, grants the interesting sense of coexistence to a collection of objects precisely when they cannot be causally related (i.e., when they are located in each other's "topological present"); such objects can be at any distance from each other.

No two views could be further apart! But both claim to represent certain intuitions about coexistence and co-presence. Gibson and Pooley sketch an error theory of the origin of such intuitions, where the idea that the present comprises objects with which we can now interact is integral to such intuitions. We mistakenly take the present thus understood to be of infinite extent, whereas, in fact, it is very large but finite. But I am inclined to regard this origin story with a grain of salt, because I do not think that intuitions about the important sense of coexistence and co-presence have a causal basis to begin with. The relevant intuitions associate the present with what exists *now*—*whether or not we can interact with it*.

---

[27] Unless spacetime itself is "gunky," i.e., does not consist of extensionless points. This option is not mentioned by Gibson and Pooley, and I set it aside here.

[28] I thus reject the following statement: "[D]uring the eleventh Apollo mission, no one in mission control in Houston could sensibly ask 'What is Armstrong doing now?' while he was on the Moon" (Gibson and Pooley 2006: 167). In my opinion, nothing could be more sensible.

Suppose Captain Kirk and Klingon Trevor are piloting two spaceships that are on course to a head-on collision (in this respect, the situation is similar to that represented in Figure 6.5). Being persons, Kirk and Trevor have subjective perspectives and, hence, are in possession of extended NOWs, perhaps determined by their "specious presents" in their respective rest frames (in that regard, the situation is different from one represented in Figure 6.5). According to AS-Coexistence, Kirk and Trevor coexist when they are located five seconds before the collision (as measured along their respective paths), but do not coexist just three seconds earlier—even though it has been true, all along, that they are headed for collision (see Figure 6.7). I take this consequence to be very counterintuitive, if not altogether absurd. CASH, of course, does not have such a consequence.

6.8. The Mixed Past of Coexistence

This is not to say that causal considerations have no role to play in the *genesis* of our concept of distant present. More plausibly, our informed concept of distant present has two "strands," causal and non-causal, which may prevail in different situations. If this is indeed the case we have a good explanation of the origin of both strands in the older worldview, where they were indistinguishable and always gave the same verdict—assuming that the older worldview incorporates the idea of instantaneous action at a distance. When it became clear that there was a limit on the speed of causal interactions the two strands came apart resulting in two distinct concepts

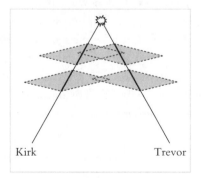

Figure 6.7. Kirk and Trevor AS-coexist five seconds before the collision of their spaceships, but not eight seconds before the collision.

of distant present which, from then on, became associated with different contexts.[29]

The situation is not unlike that with some other concepts of ordinary or scientific language that go under stress and sometimes split in two when the linguistic rules for their application come apart. The following examples are characteristic: "Are prisoners of war residents of an alien country?" "Are skis vehicles?" "Does a glass house have any windows?" "Can one dial on a push button phone?" (Sorensen 2001: 23 f). Even closer to the present context, in prerelativistic physics, the term 'mass' was used to denote both what we now call *proper mass* and what we now call *relativistic mass*.[30]

But there is an important difference between these examples and the case of distant present. In each of the above examples, the rules corresponding to the two strands of the meaning of the original term are on a par, in that they govern the application of the term in two classes of situations of the *same* general type, having to do with the use of language in various equally important contexts. The causal and the non-causal strands of the notion of distant present—those that underlie AS-Coexistence and CASH, respectively—are, however, different in the kind of their *intended* application.

The notions of AS-Present and AS-Coexistence appear to be suitable in contexts involving sentient beings who possess "specious presents," and the main significance of these notions is *subjective*. They may serve to provide a sentient being with a contextually determined psychological perspective on the spacetime world around her, but have little to say about that world itself.[31] CASH, on the other hand, is better equipped to categorize

---

[29] This particular bifurcation occurred before the advent of relativity and can be associated with the rejection of instantaneous action at a distance within the broadly classical framework. In this respect, the issue of causal connectability cuts across the classical-relativistic distinction. In particular, the items related by the classical limit of CASH are causally disconnected. But such items belong to the common classical present. Indeed, they exhaust what there is for the presentist, who will be happy to add that all such items coexist. This is not to endorse presentism, but only to indicate that causal connectability and coexistence are naturally perceived (e.g. by presentists) as being conceptually distinct and thus may go their separate ways.

[30] To use Hartry Field's 1973 example (but not his analysis of the example).

[31] Cf. Gibson and Pooley 2006: 170. They note, in particular, that in one sense, non-sentient beings do *not* have AS-Presents extending beyond their momentary location and, hence, are not capable of AS-coexisting with other objects. Gibson and Pooley then go ahead to argue that there is still a sense in which "someone *might* legitimately talk of the (non-trivial) coexistence of non-sentient objects" by attributing to them the "contextually determined duration of the *speaker's* NOW" (ibid.), as described above. The subjective nature of AS-Coexistence could not be put more clearly.

objective relations between persisting material objects populating Minkowski spacetime.

6.9. No Need in the Extended Now

I wish to end by making another brief remark about the alleged importance of the notion of the extended NOW. Some philosophers regard this notion as indispensable to the philosophy of time and related issues.[32] The initial appeal of the idea stems from the desire to bring these issues closer to experience. It appears that we do not perceive single instances of time.[33] As Gibson and Pooley note, "our basic experiential grasp of the persistence of objects involves direct perception of persistence in single, *temporally extended* experiences" (2006: 158, my emphasis).

But raw phenomena should not be confused with their theoretical analysis. This is especially important in matters of ontology. The fact that we do not experience single moments of time does not entail that they do not exist and that our experience cannot be analyzed in terms of the possession of momentary properties by objects at single instants of time. The issue has already occurred in our discussion in connection with the attribution of temporal and "lingering" properties to instantaneous object stages in stage theory (see §2.3.1). But the remarks made in that context have a more general significance and bear some amplification. As Sider notes, in a similar context:

Having [an experience] does indeed require *having had* certain features in the past. This is not inconsistent with the stage view, which interprets the past having of the relevant features as amounting to having temporal counterparts that have those features. In order to have [an experience], a stage must stand in an appropriate network of counterpart relations to other stages with appropriate features. Thus, the property *having [an experience]* is a highly relational property. It nevertheless can be instantiated by instantaneous stages.   (Sider 2001: 197–8)

---

[32] See, e.g. Arthur 2006; Savitt 2009.

[33] Or, at least, do not obviously perceive them. But what about the first instance of the light's being on, as when one enters a dark room and flips a switch? Thanks to Gilmore who suggested this example in correspondence.

To be sure, an *isolated* instantaneous stage of me could not have an experience. Neither could an enduring person existing only at an instant. But object stages are normally surrounded by other stages of that object (for the stage theorist), and enduring objects normally exist for longer than an instant (for the endurantist).[34] The upshot is that although my pain—even a sharp pain—lasts for an extended interval of time (which I may perceive as a "specious present") rather than an instant, one can legitimately attribute it to me at each moment from that interval. The same applies to the perception of music and other examples favored by the advocates of the specious present. Upon analysis, the extended NOW dissolves into a series of moments. The notion thus lacks any metaphysical implications; it has, at best, only subjective significance.

But even there its importance may sometimes be challenged. It is not so clear that even *speaking* of the *now* always requires it to have an extent of a "specious present." In some contexts, speaking of "now" appears to imply speaking of some unspecified *single* moment or, at any rate, an interval that is *shorter* than the "specious present": for example, "Go! Go!! No-ooow!!!" said of a touchdown. In other contexts, the utterance of 'now' appears to be referring to *each* moment of time in a relevant short range determined by the context of the utterance (but *not* to a "specious stretch" of such moments), as in 'Teddy is serving his sentence right now,' which means: he is doing it at *every single moment* from a relevant range of times.

Incidentally, although there is causal interaction between the speaker and the object of her talk in these cases, the interaction does not seem crucial or relevant to grounding the meaning of 'now' (Teddy may be serving his sentence on Betelgeuse, with the same effect), the point already noted above. This becomes particularly clear in contexts such as 'This star (pointing to the star) may be exploding right now, but we won't know until many years later.'

It was noted above that there is a sense in which AS-Coexistence satisfies the requirement of Objectivity. It must now be clear that this sense is rather limited. On AS-Coexistence, given two objects o_1 and o_2 and their momentary locations p_1 and p_2 in Minkowski spacetime, the fact of their coexistence is *not* determined until Δt^{NOW}_1 and Δt^{NOW}_2 are supplied, and

---

[34] A similar point could be made on behalf of the perdurantist.

these bring with them objectionable (from the ontological point of view) elements of subjectivity and context dependence. The situation becomes even more problematic when o_1 and/or o_2 are non-sentient entities. None of these problems threaten CASH.

With CASH thus rehabilitated, I now turn to more controversial issues.

Appendix 6.1:   Coexistence, Coexistence*, and Coexistence** of n objects in Galilean Spacetime

($CE_n{}^G$) Enduring or exduring objects $o_1, o_2, \ldots o_n$, q-located at points $p_1, p_2, \ldots p_n$ of ST^G respectively, *coexist* $=_{df}$ $p_1, p_2, \ldots p_n$ belong to the same moment of absolute time:

$CE_n{}^G(o_1, o_2, \ldots o_n, p_1, p_2, \ldots p_n) =_{df} t_1=t_2= \ldots =t_n.$

($CE_n{}^{G*}$) An enduring (exduring) object o_1 q-located at p_1 *coexists** with enduring (exduring) objects $o_2, o_3, \ldots o_n$, considered in abstraction from their q-locations in ST^G $=_{df}$ $o_2, o_3, \ldots o_n$ have q-locations $p_2, p_3, \ldots p_n$ respectively which, along with p_1, belong to the same moment of absolute time:

$CE_n{}^{G*}(o_1, p_1, o_2, o_3, \ldots o_n) =_{df} \exists p_2, p_3, \ldots p_n$ (o_2 is q-located at $p_2 \wedge o_3$ is q-located at $p_3 \wedge \ldots \wedge o_n$ is q-located at $p_n \wedge t_1=t_2=t_3= \ldots =t_n$).

($CE_n{}^{G**}$) Enduring (exduring) objects $o_1, o_2, \ldots o_n$ *coexist*** ("temporally overlap") in ST^G $=_{df}$ $o_1, o_2, \ldots o_n$ have q-locations that belong to the same moment of absolute time:

$CE_n{}^{G**}(o_1, o_2, \ldots o_n) =_{df} \exists p_1, p_2, \ldots p_n$ (o_1 is q-located at $p_1 \wedge o_2$ is q-located at $p_2 \wedge \ldots o_n$ is q-located at $p_n \wedge t_1=t_2= \ldots =t_n$).

($CP_n{}^G$) Diachronic parts $p_{\|1}, p_{\|2}, \ldots p_{\|n}$ of perduring objects $o_1, o_2, \ldots o_n$, respectively, *coexist* in ST^G $=_{df}$ The locations of $p_{\|1}, p_{\|2}, \ldots p_{\|n}$ belong to the same moment of absolute time:

$CP_n{}^G(p_{\|1}, p_{\|2}, \ldots p_{\|n}) =_{df} t_1=t_2= \ldots =t_n.$

(CP$_n^{G*}$) A diachronic part $p_{\|1}$ of a perduring object o_1, located at a point p_1, *coexists** with perduring objects o_2, o_3, . . . o_n in ST^G $=_{df}$ o_2, o_3, . . . o_n have diachronic parts $p_{\|2}$, $p_{\|3}$, . . . $p_{\|n}$, located at p_2, p_3, . . . p_n respectively, which, along with p_1, belong to the same moment of absolute time:

CP^{G*}($p_{\|1}$, o_2, o_3, . . . o_n) $=_{df}$ $\exists p_{\|2}$, $p_{\|3}$, . . . $p_{\|n}$ ($p_{\|2}$ is located at p_2 $\wedge$ $p_{\|3}$ is located at p_3 $\wedge$ . . . $\wedge$ $p_{\|n}$ is located at p_n $\wedge$ $t_1=t_2=t_3=$ . . . $=t_n$).

(CP$_n^{G**}$) Perduring objects o_1, o_2, . . . o_n *coexist*** ("temporally overlap") in ST^G $=_{df}$ o_1, o_2, . . . o_n have diachronic parts $p_{\|1}$, $p_{\|2}$, . . . $p_{\|n}$, located at p_1, p_2, . . . p_n respectively, which belong to the same moment of absolute time:

CP^{G**}(o_1, o_2, . . . o_n) $=_{df}$ $\exists p_{\|1}$, $p_{\|2}$, . . . $p_{\|n}$ ($p_{\|1}$ is located at p_1 $\wedge$ $p_{\|2}$ is located at p_2 $\wedge$ . . . $\wedge$ $p_{\|n}$ is located at p_n $\wedge$ $t_1=t_2=$ . . . $=t_n$).

Appendix 6.2: CASH-Coexistence, CASH-Coexistence*, and CASH-Coexistence** of *n* objects in Minkowski Spacetime

Σ^M_{CASH}(p_1, p_2, . . . p_n) $=_{df}$ $\exists F$ $t^F_1=t^F_2=$ . . . $=t^F_n$

(CE$_n^{M}$$_{CASH}$) Enduring or exduring objects o_1, o_2, . . . o_n, q-located at points p_1, p_2, . . . p_n of ST^M, respectively, *coexist* $=_{df}$ p_1, p_2, . . . p_n belong to a common time hyperplane:

CE$_n^{M}$$_{CASH}$($o_1$, o_2, . . . o_n, p_1, p_2, . . . p_n) $=_{df}$ Σ^M_{CASH}(p_1, p_2, . . . p_n).

(CE$_n^{M*}$$_{CASH}$) Enduring or exduring object o_1 q-located at p_1 *coexists** with objects o_2, o_3, . . . o_n considered in abstraction from their q-locations in ST^M $=_{df}$ o_2, o_3, . . . o_n are q-located at spacetime points p_2, p_3, . . . p_n, respectively, such that, along with p_1, they belong to a common time hyperplane:

$CO_n{}^{M*}{}_{CASH}(o_1, p_1, o_2, o_3, \ldots o_n) =_{df} \exists p_2, p_3, \ldots p_n$ (o_2 is q-located at $p_2 \wedge o_3$ is q-located at $p_3 \wedge \ldots \wedge$ o_n is q- located at $p_n \wedge \Sigma^M{}_{CASH}(p_1, p_2, \ldots p_n)$).

$(CE_n{}^{M**}{}_{CASH})$ Enduring or exduring objects $o_1, o_2, \ldots o_n$ *coexist** ("temporally overlap") in $ST^M =_{df}$ The paths of o_1, $o_2, \ldots o_n$ have points sharing a common time hyperplane:
$CE_n{}^{M**}{}_{CASH}(o_1, o_2, \ldots o_n) =_{df} \exists p_1, p_2, \ldots p_n$ (o_1 is q-located at $p_1 \wedge o_2$ is q-located at $p_2 \wedge \ldots o_n$ is q-located at $p_n \wedge \Sigma^M{}_{CASH}(p_1, p_2, \ldots p_n)$).

$(CP_n{}^{M}{}_{CASH})$ Diachronic parts $p_{\|1}, p_{\|2}, \ldots p_{\|n}$ of perduring objects o_1, $o_2, \ldots o_n$, respectively, *coexist* in $ST^M =_{df}$ The locations of $p_{\|1}, p_{\|2}, \ldots p_{\|n}$ belong to a common time hyperplane:
$CP_n{}^{M}{}_{CASH}(p_{\|1}, p_{\|2}, \ldots p_{\|n}) =_{df} \Sigma^M{}_{CASH}(p_1, p_2, \ldots p_n)$.

$(CP_n{}^{M*}{}_{CASH})$ A diachronic part $p_{\|1}$ of a perduring object o_1, located at a point p_1, *coexists** with perduring objects $o_2, o_3, \ldots o_n$ in $ST^M =_{df}$ $o_2, o_3, \ldots o_n$ have diachronic parts $p_{\|2}$, $p_{\|3}, \ldots p_{\|n}$, located at $p_2, p_3, \ldots p_n$ respectively, which, along with p_1, belong to a common time hyperplane:
$CP^{M*}{}_{CASH}(p_{\|1}, o_2, \ldots o_n) =_{df} \exists p_{\|2}, p_{\|3}, \ldots p_{\|n}$ ($p_{\|2}$ is located at $p_2 \wedge p_{\|3}$ is located at $p_3 \wedge \ldots \wedge p_{\|n}$ is located at $p_n \wedge \Sigma^M{}_{CASH}(p_1, p_2, \ldots p_n)$).

$(CP_n{}^{M**}{}_{CASH})$ Perduring objects $o_1, o_2, \ldots o_n$ *coexist** ("temporally overlap") in $ST^G =_{df} o_1, o_2, \ldots o_n$ have diachronic parts $p_{\|1}, p_{\|2}, \ldots p_{\|n}$, located at $p_1, p_2, \ldots p_n$ respectively, which belong a common time hyperplane:
$CP^{M**}{}_{CASH}(o_1, o_2, \ldots o_n) =_{df} \exists p_{\|1}, p_{\|2}, \ldots p_{\|n}$ ($p_{\|1}$ is located at $p_1 \wedge p_{\|2}$ is located at $p_2 \wedge \ldots \wedge p_{\|n}$ is located at $p_n \wedge \Sigma^M{}_{CASH}(p_1, p_2, \ldots p_n)$).

Appendix 6.3: Alexandrov-Stein Coexistence, Alexandrov-Stein Coexistence*, and Alexandrov-Stein Coexistence** of n objects in Minkowski Spacetime

$\Sigma^M{}_{AS}(p_1, p_2, \ldots p_n) =_{df} \wedge^n_{i,j=1} \Sigma^M{}_{AS}(p_i, p_j)$.

($CE_n{}^M{}_{AS}$) Enduring or exduring objects $o_1, o_2, \ldots o_n$, q-located at spacetime points $p_1, p_2, \ldots p_n$ of ST^M, respectively, *coexist* $=_{df}$ p_i and p_j are AS-co-present for all $i, j = 1, 2, \ldots n$:
$CE_n{}^M{}_{AS}(o_1, o_2, \ldots o_n, p_1, p_2, \ldots p_n) =_{df} \wedge^n_{i,j=1}$
$\Sigma^M{}_{AS}(p_i, p_j)$.

($CE_n{}^{M*}{}_A$) Enduring or exduring object o_1 q-located at spacetime point p_1 *coexists** with $o_2, o_3, \ldots o_n$ considered in abstraction from their q-locations in ST^M $=_{df}$ $o_2, o_3, \ldots o_n$ are q-located at spacetime points $p_2, p_3, \ldots p_n$, respectively, such that p_i and p_j are AS-co-present for all $i, j = 1, 2, \ldots n$:
$CE_n{}^{M*}{}_{AS}(o_1, p_1, o_2, o_3, \ldots o_n) =_{df} \exists p_2, p_3, \ldots p_n$ (o_2 is q-located at $p_2 \wedge o_3$ is q-located at $p_3 \wedge \ldots \wedge o_n$ is q-located at $p_n \wedge \wedge^n_{i,j=1} \Sigma^M{}_{AS}(p_i, p_j)$).

($CE_n{}^{M**}{}_{AS}$) Enduring or exduring objects $o_1, o_2, \ldots o_n$ *coexist** ("temporally overlap") in ST^M $=_{df}$ The paths of $o_1, o_2, \ldots o_n$ have pairwise AS-co-present points:
$CE_n{}^{M**}{}_{AS}(o_1, o_2, \ldots o_n) =_{df} \exists p_1, p_2, \ldots p_n$ (o_1 is q-located at $p_1 \wedge o_2$ is q-located at $p_2 \wedge \ldots o_n$ is q-located at $p_n \wedge \wedge^n_{i,j=1} \Sigma^M{}_{AS}(p_i, p_j)$).

($CP_n{}^M{}_{AS}$) Diachronic parts $p_{\|1}, p_{\|2}, \ldots p_{\|n}$ of perduring objects $o_1, o_2, \ldots o_n$, respectively, *coexist* in ST^M $=_{df}$ The locations of $p_{\|1}, p_{\|2}, \ldots p_{\|n}$ are pairwise AS-co-present:
$CP_n{}^M{}_{AS}(p_{\|1}, p_{\|2}, \ldots p_{\|n}) =_{df} \wedge^n_{i,j=1} \Sigma^M{}_{AS}(p_i, p_j)$.

(CP$_n^{M*}_{AS}$)   Diachronic part $p_{\|1}$ of a perduring object o_1, located at point
p_1, *coexists** with perduring objects $o_2, o_3, \ldots o_n$ in ST^M $=_{df}$
$o_2, o_3, \ldots o_n$ have diachronic parts $p_{\|2}, p_{\|3}, \ldots p_{\|n}$, whose
locations are AS–co-present with each other and with the
location of $p_{\|1}$:
CP^{M*}$_{AS}$$(p_{\|1}, o_2, o_3, \ldots o_n)$ $=_{df}$ $\exists p_{\|2}, p_{\|3}, \ldots p_{\|n}$ $(p_{\|2}$ is
located at $p_2 \wedge p_{\|3}$ is located at $p_3 \wedge \ldots \wedge p_{\|n}$ is located at
$p_n \wedge \wedge_{i,j=1}^{n} \Sigma^M{}_{AS}(p_i, p_j))$.

(CP$_n^{M**}_{AS}$)   Perduring objects $o_1, o_2, \ldots o_n$ *coexist** ("temporally over-
lap") in ST^M $=_{df}$ $o_1, o_2, \ldots o_n$ have diachronic parts $p_{\|1}$,
$p_{\|2}, \ldots p_{\|n}$, whose locations are pairwise AS–co-present:
CP^{M**}$_{AS}$$(o_1, o_2, \ldots o_n)$ $=_{df}$ $\exists p_{\|1}, p_{\|2}, \ldots p_{\|n}$ $(p_{\|1}$ is located
at $p_1 \wedge p_{\|2}$ is located at $p_2 \wedge \ldots \wedge p_{\|n}$ is located at $p_n \wedge$
$\wedge_{i,j=1}^{n} \Sigma^M{}_{AS}(p_i, p_j))$.

7

Strange Coexistence?

7.1. Coexistence and Existence@

Among the classical relations of coexistence appropriate for endurantism and exdurantism was CE^{G*}, the relation between an object q-located, or fully present, at some point and another object considered in abstraction from any of its q-locations:

> (CE^{G*}) An enduring or exduring object o_1 q-located at p_1 *coexists** with o_2 in ST^G $=_{df}$ o_2 is q-located at a point p_2 belonging to the same absolute moment of time as p_1.

In Chapter 6 we noted briefly that CE^{G*} underlies the sense in which, at any moment of o_1's career, many objects in the universe can be divided into three classes: those with which, intuitively speaking, o_1 coexists *no longer*; those with which o_1 *still* or *already* coexists; and those with which it does *not yet* coexist. As o_1 "grows older," the membership of these classes changes, thanks to the presence in our world of many objects that come to be and cease to exist. To take an example, when I was fully present on June 4, 2004 I still coexisted with the former US president Ronald Reagan (who died the next day) and already coexisted with his son, Ronald Prescott. A week later, when I was wholly present on June 11, 2004, I no longer coexisted with the former, still coexisted with the latter, and did not yet coexist with President Reagan's great-great-grandson.

These determinations can be transferred, unscathed, to Minkowski space-time, as long as we take care to track momentary spatiotemporal q-locations of a given enduring or exduring object by some useful relativistically invariant parameter, for example, the object's proper time (measuring the age of the object in its rest frame). Of course, we now have to employ the relativistic versions of the coexistence principles (i.e., CE^M_{CASH} and its derivatives).[1]

I submit that in both cases, the classical as well as relativistic, there is an interesting notion of *existence* going hand in hand with coexistence. For a given enduring or exduring object, the changing relations of coexistence it enters into during its life career provide a changing perspective on the rest of the world. When I was fully present (i.e. q-located at some moment of time) on June 4, 2004, there was a sense in which President Reagan *still existed* and George W. Bush *already* did. A week later, on the other hand, the former existed *no longer*. What reasons did I have to assert, on June 11, 2004, that George W. Bush, but not the late President Reagan, was *still in existence*? The reason seems clear: George W. Bush, but not President Reagan, coexisted with me, when I was wholly present on June 11, 2004.

Before we go any further we need to make sure that the notion of existence just introduced is sensible. It is easy to see why the notion is controversial.[2] In putting it forward, I am setting myself in opposition to one of the cornerstones of analytic philosophy, the thesis that existence is a *univocal* concept. On the eternalist view presupposed throughout this study, past, present, and future entities are equally in existence. This basic tenseless notion of existence applies to the late President Reagan and even to the Babylonian king Nebuchadnezzar no less than to George W. Bush. It would appear that adding that the latter *still* exists but the former two do so *no longer* cuts no ontological ice but merely serves to say something about the object's temporal (or spatiotemporal) location (or q-location).

But can the boundary between the ontological and the locational always be clearly drawn? I hope to provide, by the end of this chapter, enough support for the thesis[3] that the "locational" notion of existence—let us call it *existence@*, to contrast it with existence *simpliciter*—has significant

---

[1] The proper time of idealized pointlike objects is measured along their worldlines and is a well-defined notion. The situation is different with spatially extended objects, whose parts may undergo complicated relative motion. Along which spacetime trajectory should the proper time of the entire object be measured? The issue is tricky and may initially raise some doubts about the idealization of persisting objects as spatially unextended, which is adopted throughout the discussion in Chapters 6 and 7 (Gibson and Pooley 2006: 172). I address the issue in §7.9.

[2] See Gilmore 2002: 254–5. Gilmore is not alone in denying or downplaying the significance of existence@; cf. Sider 2001: 58–9. The notion of existence@ is described in more detail below.

[3] Aptly named by Gilmore the Asymmetry Thesis (Gilmore 2002: 246); I use this designation below.

implications for endurantism and exdurantism, but not for perdurantism, where its role boils down to representing certain perspectival phenomena in spacetime. This requires a lot of work, which still lies ahead. At this point I am concerned to establish the initial credentials of existence@. That by itself appears to be in defiance of common wisdom. But I think there are good reasons to explore this strategy.

Concerning the distinction between existence and existence@, note, first, that one can draw a parallel distinction between the two concepts of *co*existence, the narrow one (introduced and defended in Chapter 6), according to which I coexist with George W. Bush but not with Nebuchadnezzar, and a broader concept, according to which all inhabitants of spacetime (including me and Nebuchadnezzar) coexist with each other. The narrow concept is non-trivial and informative in a way the broad one is not. And since existence *simpliciter* goes hand in hand with the trivial notion of coexistence, it is natural to expect that there should be another notion of existence, to go along with coexistence in the interesting sense. This other notion is, of course, existence@.

Second, the distinction between existence *simpliciter* and coexistence in the trivial sense, on one side, and coexistence in the interesting sense and existence@, on the other, is similar to another, famous distinction that does a great deal of work in modal realism. As noted by David Lewis, to quantify correctly over possibilia and to do justice to modal discourse, the modal realist needs two different quantifiers: one ranging over the contents of the entire collection of possible worlds and the other restricted to a particular such world (see Lewis 1986: 3, 5–7, and elsewhere). Any two objects populating the Lewisian multi-universe coexist in the broad sense, but those belonging to different worlds fail to coexist in the restricted sense. An inhabitant of our world can state, with seriousness, that alien objects (e.g. unicorns) do not actually exist, meaning their non-existence in our world. Is such a statement ontological or "merely locational"? I suppose this is debatable. But no matter how one reads it, the statement is clearly significant.

The analogy with the temporal case works as follows.[4] Beginning with the classical context, associate with existence *simpliciter* the unrestricted

---

[4] The modal-temporal analogies have been extensively debated by philosophers of time since Arthur Prior's seminal work (Prior 1957). Thus many presentists have drawn parallels between non-present times and modal actualism's treatment of non-actual worlds, insisting that both types of entities fail to exist. Many eternalists have endorsed the parallel but turned it around: all moments of time are equally

quantifier ranging over all enduring or exduring objects.[5] Now for each particular t held fixed, introduce a restricted quantifier ranging over objects q-located at t (i.e. at spacetime points belonging to the absolute t-hyperplane). The meaning of 'existence@' is then associated with the family of such restricted quantifiers.

Consider two enduring or exduring objects o_1 and o_2 q-located at t. Such "t-mates" coexist both in the trivial and in the interesting sense. Consequently, for o_1 at t, o_2 coexists with it (hence, exists@) *still* or *already*. Suppose another enduring object o_3 ends its career at $t' < t$. Then although o_1 at t and o_3 at t' coexist in the trivial sense, they fail to be t-mates. For o_1 at t, o_3 *no longer* coexists with it, hence no longer exists@.

One might object that these locutions betray existence@ as an explicit relation to time and 'o exists@' as an incomplete expression, and that this is inappropriate to any concept of existence deserving the name. To this, I reply that the concept of existence at a world corresponding to the Lewisian restricted quantifier can be similarly looked upon as a relation (namely, the world-mate relation) to some inhabitant of a given world and the corresponding expression 'o exists' (meaning its existence in that world) as an incomplete expression. If that is unproblematic (which I think it is), then I do not see why the relational (in a similar sense) nature of existence@ should be problematic. In essence, existence@ behaves like the indexical notion of actuality does in modal realism. Any object that is a world-mate of a given object is actual, where this is meant to have the same force as saying that every world is actual at itself. Similarly, for any enduring or exduring object q-located at t, another such object exists@ just in case it is also q-located at t. Statements of actuality in modal realism go hand in hand

real, just as all worlds are in the ontology of modal realism. More tendentiously, the issue is sometimes put in a slogan: according to presentism, time is (in many ways) like modality; according to eternalism, time is (in many ways) like space. This, of course, begs the question against the view that modality is (at least in some sense) like space! The analogies between time and modality have their limitations, and one may be inclined to reject them altogether; see Meyer 2006 for a particularly critical approach. On the other hand, certain features of the analogies may suggest attractive package deals; e.g. the counterpart analysis of possibility can be combined with the counterpart analysis of temporal predication, as is done in Sider 2001. My intentions below are far from defending modal realism. All I want to do is to exploit one of its features to substantiate the distinctions I am interested in—(a) that between the interesting and the trivial notions of coexistence in spacetime and (b) the related distinction between existence and existence@.

[5] This presupposes that enduring objects related by identity and exduring objects related by genidentity should be counted only once. How can counting be other than by identity? I set this contentious issue aside. For recent discussions see Sider 2001: 188–93 and Hawley 2001: 62–4.

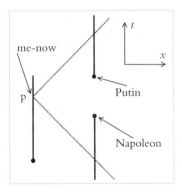

Figure 7.1. I *still* coexist with Napoleon, and I *already* coexist with Putin. Both are "temporally here" for me now.

with statements of restricted coexistence, except that the former suppress mention of the object with respect to which actuality is covertly relative, via the world-mate relation. Similarly, statements of existence@ go hand in hand with statements of temporally restricted coexistence, except that the former suppress mention of the pair consisting of an object and a time (or an achronal spacetime region) with respect to which existence@ is covertly relative, via the t-mate relation.

In light of CASH,[6] the relativistic generalization of existence@ requires replacing the relation of belonging to the same moment of absolute time with the relation of belonging to the same hyperplane of simultaneity. If I am wholly present (i.e. q-located) here and now (at p), any other enduring or exduring object exists@ just in case it coexists with me now (i.e. is wholly present at p* spacelike separated from p). Moreover, it is appropriate to qualify the existence@ of such objects with temporal modifiers *still*, *already*, *not yet*, and the like. It is especially appropriate when the objects in question come to be and cease to exist "during" my own life career.

This gives rise to the following situation represented in Figure 7.1.[7] Suppose I am wholly present at a certain moment of my proper time far away in the Milky Way. Then I *still* coexist with Napoleon and I *already* coexist with Vladimir Putin. Moreover, in view of the connection between coexistence and existence@, Napoleon still exists@, and Putin already does.

---

[6] Coexistence As Sharing a Hyperplane of simultaneity. See §6.4.
[7] The essential features of the situation come from Balashov 2000*a*: 156 and 2000*b*: S560.

In Gilmore's apt expression (2002: 245), they are both "temporally here" for me now. In Balashov 2000*a*, 2000*b* I argued that this is unacceptable. There is no *temporally laden* sense in which Napoleon and Putin can be in existence or "temporally here" together: the former's end lies in the absolute past of the latter's beginning. Since endurantists and exdurantists are committed to this temporally laden "togetherness" claim in the relativistic context, endurantism and exdurantism are inferior to perdurantism.

   This is intended as an opening salvo. One way to embellish the strategy (and to encourage patience), which was suggested by Gilmore (2002), is to identify two separate claims that come together to produce the above conclusion, the Asymmetry Thesis and the Absurdity Thesis. According to the former, the endurantist and the exdurantist are, but the perdurantist is not committed to the temporally laden (and hence, potentially threatening) reading of the togetherness claim. According to the Absurdity Thesis, this commitment is harmful to endurantism and exdurantism. Gilmore objects to both theses.[8] In the next section I respond to his objections to the Asymmetry Thesis. I then grant his observation that resisting those objections undermines the Absurdity Thesis, which must, in the end, be abandoned in its original form. This, however, opens a way to other closely related relativistic arguments against endurantism and exdurantism that are immune to Gilmore's critique. I develop two such arguments in §§7.7 and 7.8 and respond to further objections due to Gibson and Pooley 2006 (§7.9). All along I make use of the material of Chapters 5 and 6 and of the temporal–modal analogy developed above.

7.2. The Asymmetry Thesis

What makes Napoleon and Putin "temporally here" for me now (see Figure 7.1)—when I am q-located at p—is that, when I am so q-located, I coexist with both of them in one of the derivative senses of the basic coexistence relation CE^M_{CASH} appropriate for endurance and exdurance. The principle governing this derivative relation (i.e. coexistence*) was introduced in §6.4:

---

[8] Gilmore stated his objections in the context of criticizing my earlier relativistic argument against endurance, and endurance alone. Here I take the liberty to extend the earlier argument, as well as Gilmore's critique, to exdurance, as well as endurance.

($CE_n{}^{M*}{}_{CASH}$) Enduring or exduring object o_1 q-located at spacetime point p_1 *coexists** with o_2 considered in abstraction from its q-location in ST^M $=_{df}$ o_2 is q-located at a spacetime point that shares a common time hyperplane with p_1.

But recall that the corresponding perdurantist principle is easily available too:

($CP_n{}^{M*}{}_{CASH}$) A diachronic part $p_{\|1}$ of a perduring object o_1, located at point p_1, *coexists** with another perduring object o_2 in ST^M $=_{df}$ o_2 has a diachronic part $p_{\|2}$ located at point p_2, which shares a common time hyperplane with p_1.

This suggests that the perdurantist may also be entitled to an interesting notion of existence—call it existence^P@—similar to the endurantist/exdurantist notion of existence@. In particular, the perdurantist may want to say that, for any diachronic part $p_\|$ of a perduring object o, located at p, another perduring object exists^P@ just in case it has a diachronic part located at a spacetime point that is spacelike separated from p. And this implies that in scenarios such as one involving Napoleon and Putin, the perdurantist is committed to an analog of the "togetherness claim." Suppose I am a perduring object and one of my diachronic parts is located at a certain moment t of my proper time somewhere in the Milky Way. Call this part Yuripart. According to $CP^{M*}{}_{CASH}$, Yuripart coexists* with Napoleon and it also coexists* with Putin. Moreover, in view of the connection between coexistence* and existence^P@, Napolean exists^P@ and so does Putin. They are both "temporally here" for Yuripart. It would seem that if the togetherness claim is unacceptable for the endurantist and the exdurantist, then the perdurantist is no better off.

My reasons for thinking otherwise are based on two distinctive common features of endurance and exdurance associated with multi-q-location. The first such feature is that (a) enduring and exduring objects can be said to be wholly present at each moment of their proper time. My full presence at a certain place and time is, according to this notion, the focal point of my connection with the rest of the universe. (b) Relatedly, enduring and exduring objects can be said to change their q-location with their proper time.[9] This induces a corresponding change in relations of such objects

---

[9] In the sense in which an object's q-location can be regarded as a function of its proper time.

to their environment. As I become older, the membership of the class of objects with which I coexist*—hence of those that exist@—undergoes change. This legitimizes ascribing to my coexistence* with them, and hence to their existence@, temporally laden determinations, such as *still*, *already*, and the like. I *used* to coexist (i.e. to coexist*) with President Reagan when I was fully present on June 4, 2004 at my usual location. He *still* existed (i.e. existed@) then. And George W. Bush existed@ then *already*. I *continue* to coexist with the latter but *no longer* with the former. The late President Reagan *no longer* exists (i.e. no longer exists@). As long as such qualifications are legitimate, they lead the endurantist and the exdurantist to the togetherness claim, which, I have contended, is unacceptable.

There is no place for anything like (a) and (b) in the perdurantist ontology.[10] If I am a perduring object, I am never fully present at any one place and time. At any such location, I am present only partially, and being only partially present there does not entitle me to temporally laden determinations, as regards the existence@ of other objects and their coexistence* with me, that being fully present does. Moreover, there is no sense in which perduring entities *change* their location in spacetime, no strict sense in which they (considered in their entirety) *age*, and thus no reasonable sense in which their coexistence* with other objects (and hence those objects' existence@) can be characterized in terms of changing temporally laden qualifications (*still*, *already*, *no longer*, etc.). The reason perduring objects cannot do any of that is that they are not capable of multilocation (or multi-q-location).

Absent such temporally laden connotations, the perdurantist notions of coexistence* and existence^P@ give rise, in the case of Napoleon and Putin, to the following innocuous situation. My diachronic part referred to above as Yuripart, Napoleon's last diachronic part, and Putin's first diachronic part are all confined to their locations in Minkowski spacetime, and the fact that Yuripart coexists* with both is simply a further tenseless fact about us and our parts. The perdurantist version of the togetherness claim is therefore entirely harmless. This establishes the Asymmetry Thesis.

---

[10] I note a helping hand from Gibson and Pooley 2006: 173, who endorse this contrast, *pace* Gilmore.

To recap, being wholly present at a spacetime point determines an enduring or exduring object's changing relations to the rest of the universe in a way being only partially present at such a point does not for a perduring object. This prompts the endurantist and the exdurantist to grant a temporally laden sense of existence@ to objects that cannot be in existence together in this sense, such as Napoleon and Putin. The perdurantist avoids a commitment to this conclusion because partial presence at a spacetime point does not invest a corresponding perduring object with a determinate temporary relation to the world and, hence, does not warrant ascribing temporally laden determinations, such as *still* and *already*, to the existence of other objects.

But Gilmore has pressed the charge that although one cannot expect this service from a total four-dimensional perduring object, it could still be performed by its diachronic part. Such a part is wholly present at a single spacetime point. Does not this allow one to associate with it a definite perspective on the rest of the world, thus restoring parity with endurantism and exdurantism? If it does then, if the endurantist/exdurantist version of the togetherness claim is offending, the perdurantist version of it fares no better. Given CP^{M*}_{CASH}, my diachronic part (i.e., Yuripart) coexists* with both Napoleon and Putin. Given the connection between coexistence* and existence^P@, they both exist^P@ (for Yuripart), which is just as unacceptable (or just as acceptable, as the case might be) as the corresponding claim on behalf of the endurantist and the exdurantist. Moreover, there is a sense in which the perdurantist claim is also temporally laden. As Gilmore notes, "Insofar as perdurantism can be taken seriously, it must be consistent with . . . [the] undeniable fact that there is *at least some weak sense* in which George W. Bush is *still* or *already* in existence for me at the current point on my worldline" (2002: 252).

But is this *weak sense* strong enough to yield a togetherness claim as offending as the corresponding endurantist or exdurantist claims are? I do not think so. The perdurantist claim is grounded in the perspective of a momentary diachronic part of a perduring object, such as Yuripart. However, in the perdurantist ontology, Yuripart is not me but only a short stage of me. So what is wholly present at a spacetime point—and what has a definite relation to the rest of the world—is not an ordinary object, Yuri, but a less familiar entity, Yuripart. In contrast, what is the focal point of such a relation in the cases of endurantism and exdurantism is the ordinary person in its entirety. The contrast is quintessential to the whole issue between endurantism and exdurantism, on the one hand, and perdurantism, on the other. I submit that it cuts deeply enough to render the perdurantist version of the togetherness claim metaphysically innocent. The "weak sense," which is at work in Gilmore's objection, is the *vicarious* sense in which the properties of diachronic parts of perduring objects can be attributed to the four-dimensional wholes. Strictly speaking, Yuripart represents *my* perspective on the world no more than Yuripart$_1$, Yuripart$_2$, and countless other diachronic parts of me do and thus cannot bear the weight of temporally laden determinations, such as *still* and *already*, pertaining to the existence of other objects. Such an object may be already in existence for Yuripart$_1$ but not yet for Yuripart$_7$. And since neither Yuripart$_1$, nor Yuripart$_7$, nor any other diachronic part of me represents *my* perspective par excellence, it is unclear what we ought to say about the existence (i.e. existence^P@) of the given object. On the other hand, if I am an enduring or exduring object wholly present at *t*, then my relation to other objects, associated with my temporary location, is determinate and pertains to me in my entirety.

This underscores the extent to which the ontology of perdurantism is revisionary. It denies that objects have temporary properties in anything stronger than a "secondary" sense. Indeed, it is this denial that allows perdurantism to avoid the problem of coincident objects ("temporal overlap is not coincidence"), which constitutes one of the reasons in its favor.[11]

This concludes my defense of the Asymmetry Thesis. In conjunction with CASH, it commits endurantists and exdurantists to the temporally laden version of the togetherness claim, which was briefly described above. To repeat, if I am wholly present at a certain far-away point, I coexist* with Napoleon and also coexist* with Putin. This authorizes me to say that they both exist@—both are "temporally here" for me. And this, I have suggested, is unacceptable. Napoleon and Putin cannot be in existence together in any temporally laden sense. Claims of this sort are absurd. But Gilmore has argued that they are not. Let us turn to this argument.

---

[11] See Heller 2000 for a recent discussion.

7.3. The Absurdity Thesis

More precisely, Gilmore has argued that anyone who subscribes to CASH should not be offended by the temporally laden version of the togetherness claim. CASH by itself involves such a radical departure from our naïve intuitive ideas about coexistence that making a further step by accepting the consequences of the togetherness claim should not be especially troubling. Thus Gilmore:

If I am willing to broaden my ideas about coexistence so as to allow for the possibility that I now coexist with both of two things that never coexist with each other, then I should also be willing to broaden my ideas associated with the phrase "are still or already in existence for me": I should then be willing to broaden these latter ideas so as to allow for the possibility that there is at least some weak sense in which the given phrase can apply to both of two things that never coexist with each other. (2002: 246)

Is there room for disagreement here? The whole issue boils down, at this point, to the question of how to extend our ordinary notions into the relativistic domain: what can be sacrificed and what has to be retained at all costs. And since we have already left behind many ordinary beliefs (e.g. those about absolute simultaneity and universal time), we are swimming in uncharted waters. Consequently, there may be no shared standards that could warrant extrapolation of old concepts to the new domain to everyone's satisfaction. One might still insist that endurantists and exdurantists must accept CASH, simply because it is the only plausible extension of the interesting sense of coexistence into the relativistic domain, and, at the same time, reject the temporally laden togetherness claim. But instead of pursuing this line, let me concede that the case is not conclusive. Having defended the Asymmetry Thesis, I am inclined to agree with Gilmore that endorsing CASH takes the sting from the endurantist/exdurantist version of the togetherness claim.

But the foregoing analysis suggests a way of reinforcing the case. Building on the results of Chapter 6, I develop, in the remainder of this chapter, two other arguments of the same broad variety (i.e. also centered on the notion of coexistence), which show that the togetherness claim, even if relatively innocent by itself, is just one of a family of progressively more harmful consequences of the general idea that coexistence in Minkowski spacetime

...

hinges on sharing a hyperplane of simultaneity, and that the pressure on the endurantist and the exdurantist to reject those other consequences is greater than the pressure, defied by Gilmore (and, I suppose, by others[12]), to reject the simple togetherness claim.

The precursors of the arguments of the following sections were first stated in the context of exdurantism, not endurantism (Balashov 2002). The discussion below applies to both views, as I continue to rely on their common feature: multi-q-location. The new arguments make essential use of the fact that interesting coexistence in spacetime is a *multigrade* relation (see §6.2 and 6.4).

7.4. Collective CASH Value of Coexistence

It should be noted, in this connection, that the potentially offending version of the togetherness claim considered above was predicated on the overly restricted use of CASH, as applied to a pair of objects. Indeed, the situation involves an enduring or exduring object o, wholly present at spacetime point p, and two other such objects, o_1 and o_2, considered in abstraction from their q-locations, such that o at p coexists* with o_1, and it also coexists* with o_2. This entails that o_1 and o_2 are both in existence@ (for o at p). But o_1 does not coexist with o_2 in any of the senses allowed by the endurantist and exdurantist principles of coexistence. As we have learned by now, this may not be "absurd." I want to suggest that whatever trouble there may be in this situation, it should be blamed primarily on the restricted *grade* of the coexistence relations involved in this example.

Given that sharing a hyperplane of simultaneity (HPS) is not transitive and that coexistence holds between pairs of objects, should someone who is committed to CASH be surprised to discover that in some cases, o at p coexists* with o_1 and also coexists* with o_2, but o_1 does not coexist* with o_2? Not necessarily. After all, from the fact that o at p coexists* with o_1 and o at p coexists* with o_2, it does not even follow that o at p coexists* *with o_1 and o_2*. True, the latter can be said to be both in existence (i.e. in existence@) for o at p—but, one wants to add, "not in the same mode."

---

[12] Unless I misinterpret Kristie Miller's intentions, she may be getting at the same point in her brief critical comment on my earlier argument (Miller 2004: 360–1).

Their existence@ has different (explicit or covert) indexical modifiers. o at p shares a certain HPS with o_1 (when the latter is wholly present at a suitable spacetime point), and it shares another such HPS with o_2. But there is no single HPS that they all share *collectively*. If there were such a common HPS then the situation that led to the allegedly offending togetherness claim would never arise in the first place and, hence, endurantists and exdurantists would have nothing to worry about.

These observations suggest that, given non-transitivity of the grounding relation of sharing a common HPS, it may be unfair to saddle the endurantist or the exdurantist with the dyadic relation of coexistence and then go on to blame her, in essence, for the non-transitive character of this relation. But we know (discussed in §§6.2 and 6.4) that grounding the interesting notion of coexistence (i.e., one that is neither empty nor universal) in a relation of fixed grade is not the best way to treat this notion to begin with. Things must be capable of coexisting, not merely pairwise, but *en masse*.

What is needed, therefore, is simply upgrading the relation $CASH_2$, the two-grade relation of sharing an HPS, at work in the above example (see Figure 7.1), to $CASH_n$, the n-grade relation of sharing a *single* HPS, using the principles developed in §6.4 (and explicitly stated in Appendix 6.2): $CE_n^M{}_{CASH}$, $CE_n^{M*}{}_{CASH}$, $CE_n^{M**}{}_{CASH}$, and their perdurantist counterparts. We have seen that $CASH_n$ is an improvement on $CASH_2$. It allows one to accommodate the idea that many objects—all the people currently alive, all the planets in the Solar system, all the *Star Trek* characters—can stand in a single relation of coexistence to each other. It is true that if n momentary spacetime q-locations (of enduring or exduring objects, or of temporal parts of perduring objects) satisfy $CASH_n$, then taken pairwise, they also satisfy $CASS_2$. But as already noted,[13] the contrary does not hold. In this respect, $CASH_n$ is more comprehensive than $CASS_2$. It grants coexistence *en masse* to a family of objects that, intuitively speaking, coexist all together and not just pairwise. To reiterate, the collective CASH value of coexistence in ST^M is greater than the sum of its piecemeal values.

$CASH_n$ also eliminates the injustice, which may (as I have conceded) have been done to the endurantist and the exdurantist by the Absurdity Thesis, and, hence, it eliminates the reason for Gilmore's objection to

---

[13] §6.4; see especially n. 12.

this thesis, by precluding the "absurd" situations from arising in the first place. If I am wholly present at some moment of my proper time on Betelgeuse, then I coexist* with Putin and I also coexist* with Napoleon in the sense of $CE^{M*}{}_{CASH}$, but *not* in the sense of $CE_n{}^{M*}{}_{CASH}$. And this makes it entirely perspicuous that broadening our ideas about coexistence, in making a transition to the relativistic domain, must go hand in hand, as Gilmore has suggested (2002: 246), with broadening our ideas about the use of temporal determinations such as *still* and *already*. On $CE_n{}^{M*}{}_{CASH}$, it is simply not the case that I still coexist with Napoleon and already coexist with Putin. But this is so because I do not coexist* with them *at all*: we do not share a common HPS. This blocks the next move, the ascription of the temporally laden sense of existence@ to both of them, at an early stage.

The classical limit of $CASH_n$ is the old classical notion of existing at the same moment of absolute time. This is also true of $CASH_2$. But in the case of $CASH_n$, recovering an intuitively correct classical limit is especially valuable, since $CASH_n$ is broader in scope and thus comes closer to the common notion of coexistence as a multigrade relation, even if it becomes, as a result, less permissive in granting mutual coexistence to a collection of objects.

Our primary interest, however, lies in those features of $CASH_n$-based coexistence relations in which they deviate from their classical counterparts. Before turning to them, let us revisit the important notion of existence@ and introduce its multigrade version. It will be convenient to return briefly to the classical (non-relativistic) context. This may also be a good occasion to switch to different examples.

7.5. Collective Existence@ and Coexistence in Classical Spacetime

There is a familiar sense in which the enduring Descartes, Galileo, Kepler, and Brahe coexisted in 1600, but not in 1620 (see Figure 7.2). And this sense is open to the exdurantist too. The q-locations of these persons at some point in 1600 belong to the same moment of absolute time. This is not true of any moment in 1620, because (sadly) Brahe has no q-location at any moment in 1620. We shall continue to call items sharing a common

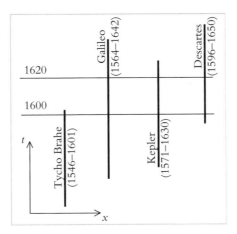

Figure 7.2. Descartes, Galileo, Kepler, and Brahe classically coexist in 1600, but not in 1620.

moment *t* of absolute time *t*-mates. For the endurantist and the exdurantist, such items are ordinary objects wholly present at *t*. For the perdurantist, they are momentary *t*-parts of ordinary perduring objects.

The enduring or exduring Descartes, Galileo, Kepler, and Brahe are 1600-mates but not 1620-mates. In the classical context, entities that are *t*-mates coexist (*en masse*) in the interesting sense and exist@ in the sense associated with the restricted quantifier ranging over items wholly present at *t*. On the other hand, all objects whatsoever—past, present, and future—coexist in the trivial sense, and, of course, all such objects exist (*simpliciter*).

In particular, Galileo, Kepler, and Brahe all exist@ for Descartes at 1600, but not for Descartes at 1620. All four coexist, in the interesting sense, in 1600, but not in 1620. On the other hand, all these persons also coexist with each other in the trivial sense, by dint of populating the same 4D spacetime manifold and, of course, they coexist in that sense with everything else that ever existed, exists, or will exist in the entire history of the universe.

Just like their dyadic predecessors, the multigrade endurantist and exdurantist notions of coexistence and existence@ are temporally laden. It makes sense to say that, in 1600, Galileo *still* coexisted with Brahe and *already* coexisted with Descartes and Kepler. Galileo *continued* to coexist with Descartes and Kepler in 1620, but *no longer* with Brahe. As Galileo grew older, his coexistence relations with other objects in the universe underwent change and, with them, his perspective on their existence. It is perfectly reasonable to say that, for Galileo in 1600, Descartes, Kepler, and Brahe were still in existence (i.e. in existence@). This is no longer the case for Galileo in 1620.

To sum up, if objects endure or exdure then besides the trivial sense, in which all of them share a single four-dimensional spacetime world, there is an interesting sense in which they collectively share or fail to share "temporal worlds" at particular moments of their individual time (which, in the prerelativistic setting, coincides with the common absolute time), just as there is an interesting sense in which Lewisian objects may or may not be world-mates, despite the fact that there is a broader sense in which they all coexist by sharing the entire Lewisian pluriverse. The analogy should by now be clear.

There is, however, a notable disanalogy between the worlds of modal realism and classical temporal worlds, which can be associated with endurance and, in a less strict sense, with exdurance. No two Lewisian possible worlds overlap, because no object can exist in more than one such world. But if endurantism is true then distinct temporal worlds (i.e., the occupants of distinct absolute hyperplanes of simultaneity) may contain literally the same objects. Thus the 1600-world and the 1620-world both contain the selfsame Descartes, Galileo, Kepler, and many other enduring objects populating both worlds. If exdurantism is true, then this can be said with some qualification, which amounts to replacing the sameness relation construed as strict identity with the sameness relation construed as genidentity.[14]

On the other hand, if perdurantism is true, then distinct classical temporal worlds do not overlap, in the sense of containing the same *ordinary* objects. In this respect, the classical ontology of perdurance shares more common features with modal realism than classical endurance and exdurance. But it is a consequence of this affinity that the corresponding perdurantist notions of coexistence (and of existence^P@) are devoid of temporally laden connotations. There is no sufficiently strong sense in which the perduring Descartes becomes older and his view on the rest of the world changes.

---

[14] On the exdurantist notion of sameness as distinct from identity, see §2.3 and Hawley 2001: 62–4. See also n. 5 above.

This, of course, is simply more grist for the Asymmetry Thesis (see §7.2), now generalized in accordance with CASH$_n$.

Let me now introduce the relativistic versions of the above notions.

7.6. Collective Existence@ and Coexistence in Minkowski Spacetime

According to CASH, coexistence in Minkowski spacetime is a matter of sharing a hyperplane of simultaneity (HPS), as reflected in the statements of $CE_n{}^M{}_{CASH}$, $CE_n{}^{M*}{}_{CASH}$, $CE_n{}^{M**}{}_{CASH}$, and their perdurantist counterparts (discussed in §6.4 and Appendix 6.2). In difference from the classical case, there is no privileged family of HPSs. But the notion of sharing a common HPS is as close as one can get to the classical notion of sharing a moment of common time. In fact, an HPS represents a moment of time in a particular frame of reference. Just like its classical counterpart, it stretches across the entire world. Indeed, the occupant of a given HPS *is* the world at a certain moment of time in some inertial frame. Let us refer to it as a *temporal-like world*, and let us call the inhabitants of a temporal-like world *HPS-mates*. Such inhabitants may be enduring or exduring objects wholly present at spacetime points lying on a given HPS, or momentary temporal parts of perduring objects. HPS-mates all coexist with each other in an interesting sense. And for any entity wholly present at a spacetime point, there is a sense in which its HPS-mates exist@. But in light of the Asymmetry Thesis, the interesting notion of coexistence and the corresponding notion of existence@ are temporally laden only for the endurantist and the exdurantist, and not for the perdurantist. (More on this in the next two sections.)

For all practical purposes, the relativistic notions based on CASH are indistinguishable from their classical limits when they are applied to objects of everyday life. Thus, what has just been said about Descartes and his contemporaries remains valid in the relativistic framework. In general, an inquiry about what enduring or exduring objects coexist with what other such objects in relativistic spacetime should start with a particular object of interest *o*, wholly present at spacetime point p at a certain time *t* in its life career (i.e. its proper time), and then pose the question as to what objects it coexists with (in the sense of $CE_n{}^M{}_{CASH}$). Other objects enter into such a relation of coexistence with *o* at *t* by being wholly present at spacetime points sharing a common HPS with p. On this approach, *o* at p turns out to coexist (as it should) with what we would prerelativistically count as its "contemporaries" and not coexist with any of its "predecessors."

The notion of a relativistic "contemporary" is, of course, different from its classical counterpart. Classical contemporaries exist at the same moment of a single time, the absolute Newtonian time. No such concept is available in the relativistic framework. But there is a good substitute. Each relativistic contemporary of *o* at *t* is a *certain age*, the age in question being measured by the proper time of that contemporary.[15] In the end, this enables all relativistic contemporaries, including *o* itself, to enter into a single many-place relation of coexistence with each other on a par. Thus, the 30-year-old Data, the 46-year-old Captain James T. Kirk, and the 65-year-old Captain Jean Luc Picard are contemporaries of each other; they are HPS-mates. On the other hand, Klingon Trevor is the relativistic predecessor of all of them: none of his momentary locations shares a temporal-like world with any momentary location of the first three (see Figure 7.3).

Notably, the same 30-year-old Data is also a contemporary of the 40-year-old Captain Kirk and the 55-year-old Captain Picard. This is a

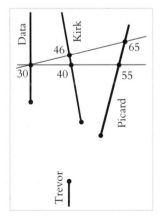

Figure 7.3. Relativistic contemporaries and successors.

---

[15] Is the notion of proper time well defined for spatially extended objects such as Descartes and Kepler? If not, doesn't it threaten the idealization of persisting objects as spatially unextended, which is at work in the arguments of Chapters 6 and 7? The issue, which was brought to light by Gibson and Pooley 2006: 172, is discussed at the end of this chapter.

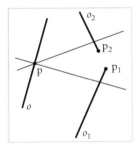

Figure 7.4. o at p coexists* with o_1 and with o_2, but not with both; o_2 is entirely in the absolute future of o_1.

consequence of the latitude allowed by relativistic spacetime. It may be surprising but it does not look pernicious. Not yet! But the point to note here is that, unlike classical temporal worlds, their relativistic descendants may "crisscross": two HPSs may intersect "at an angle" in spacetime and thus contain the same enduring or exduring object of a certain age or the same stage of a perduring object. This is in addition to the sense in which even classical temporal worlds, which do not literally intersect in spacetime (since the absolute HPSs they populate are always parallel), may nonetheless overlap in virtue of containing the same enduring object (or the same exduring object, with 'same' interpreted in a way appropriate for exdurance).

7.7. Contextuality

Let us return for a moment to the situation discussed in connection with the Absurdity Thesis. An enduring or exduring object o at p coexists*, in the sense of CE^{M*}_{CASH}, with o_1 and with o_2 but not with both: there is no single temporal-like world containing p and some momentary q-locations of both o_1 and o_2 (see Figure 7.4). Informed by Gilmore's reasons against the Absurdity Thesis, we may wonder, however, why there should be such a single world, given that o_2 lies entirely in the absolute future of o_1 and that, consequently, o_1 and o_2 fail to coexist with each other. o_1 may be Captain Kirk's great-grandfather and o_2 his great-grandson. It would be rather strange for anyone to be a "contemporary" of both.

So far so good. But the problem is easily reproduced by separating o_1 and o_2 widely enough in space (see Figure 7.5). There we have a situation

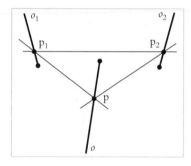

Figure 7.5. o at p coexists* with o_1 and with o_2, but not with both; in addition, o_1 at p_1 coexists with o_2 at p_2.

in which o_1 *does* coexist with o_2 (their momentary locations p_1 and p_2 share a common HPS), and o at p coexists* with o_1 and with o_2, but *not with both*. A similar feature is characteristic of the notion of relativistic *overlap*, applying to enduring or exduring objects considered in abstraction from their particular q-locations in spacetime:

$(CE_n{}^{M**}_{CASH})$ Enduring or exduring objects o_1, o_2, . . . o_n *coexist*** ("temporally overlap") in ST^M $=_{df}$ The paths of o_1, o_2, . . . o_n have points sharing a common time hyperplane.[16]

Consider three enduring or exduring objects that come to be and cease to exist. Then they may coexist pairwise *at some point or other* of their life careers. In other words, they may overlap pairwise. But this does not guarantee that they coexist (in the same sense) *all together*. The failure of such mutual coexistence occurs in cases in which no selection of momentary locations of the three objects can be unified by a single triadic HPS–mate relation. Thus it may be the case that Data overlaps with Captain Kirk, Captain Kirk overlaps with Captain Picard, and the latter with Data. Taken pairwise, they all share temporal-like worlds with each other. Taken all together, however, they do not share any single temporal-like world (see Figure 7.6).

The same is true of a quadruple of objects. All the triples constructed out of them may coexist, in the sense of $CE_n{}^{M**}_{CASH}$ or overlap; but this

---

[16] "Relativistic overlap" governed by $CE_n{}^{M**}_{CASH}$ is a further generalization of coexistence* (§6.4); it was stated earlier in Appendix 6.2.

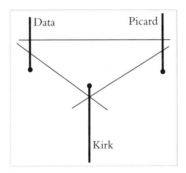

Figure 7.6. Kirk, Picard, and Data coexist** pairwise, but not all together.

does not entail the coexistence or overlap of all four. (I shall not attempt to illustrate such a case in a diagram.)

These results, which can be extended to more numerous collections of objects, show that coexistence (i.e., coexistence* and coexistence**) in the relativistic world of enduring, or exduring objects, is partitioned in a most peculiar way bearing the mark of contextuality: facts about coexistence of the members of a collection of objects, however numerous, become sensitive to what other objects are taken into account. Such facts do not "add up" properly.

Contextuality of coexistence should not be confused with the breakdown of transitivity. As we know by now, the various relations of coexistence between members of *pairs* of objects in the relativistic world generally fail to be transitive. For example, coexistence, in the sense of overlap, of o_1 with o_2 and of o_1 with o_3 does not entail coexistence, in the same sense, of o_2 with o_3. Contextuality, on the other hand, means that coexistence between the members of *all* such pairs does not entail coexistence among the members of the whole *triple*: o_1, o_2 and o_3.

Why should anyone be bothered by contextuality but not so much by the lack of transitivity? One reason (already noted in §6.4) is that transitivity fails, for some senses of coexistence, even in the classical situation. I coexist** (i.e. temporally overlap) with my father and he coexists** (in the same sense) with my grandfather. But I do not coexist** with my grandfather. However, the classical relation of coexistence is free of contextuality. If my father, my grandfather, and I overlap pairwise then we overlap all together. We do not find this remarkable because we simply take it for granted. But for endurantists and exdurantists who take relativity seriously, something important is at stake here.

Indeed, if objects endure or exdure then they live in temporal-like worlds; those are where they are wholly present and those are what they share—much as objects in Lewis's ontology live in and share particular possible worlds. In a trivial sense, the latter also share the entire collection of worlds. But nothing significant turns on this broader notion. All the important features of the possible worlds ontology, including the modal properties of objects, are grounded in the facts about what objects belong to what worlds. The fact that every object also belongs to the whole collection of worlds bakes no bread. Similarly, endurantists and exdurantists should take the facts about what object belongs to what temporal-like world at what point in its career—and what other objects it shares that world with—as the ground of all the important features exhibited in the temporal "multi-universe." These include temporary properties of objects and their changing relations with each other. The fact that all objects trivially share the single spacetime manifold is simply too generic to underwrite any interesting relations of that sort.

Both endurantists and exdurantists must thus recognize existence in a temporal (or temporal-like) world and sharing such a world as basic facts. But then it is natural to expect such facts to obey a reasonable "calculus." And they certainly do so in the classical case, where the t-mate relation is grounded in the absolute simultaneity among momentary q-locations of enduring and exduring objects. Minkowski spacetime suggests a bona fide candidate to do a similar job, the HPS-mate relation, which adequately recovers its prerelativistic counterpart in the classical limit. The problem with it is that it stumbles upon a simple rule that is, intuitively, part and parcel of the concept of coexistence. To put it in a grotesque form, if *any* 999 members of a collection of 1000 objects coexist with each other, then *all* 1000 objects must do so as well.

But isn't the same true of momentary parts, or stages, of perduring objects? After all, aren't they similar to enduring objects (and, even more so, to exduring ones) in this respect at least that they are wholly present at spacetime points? If so, the stages' coexistence relations with each other in the relativistic world must be partitioned in the same way as the corresponding relations of enduring or exduring objects taken at particular moments of their careers. And if the latter fail to obey a "reasonable calculus" then so must the former.

I cannot but agree with this. And I take this to be a reason to prefer the perdurance theory of persistence both to the endurance and exdurance theories. It has already emerged in our discussion of the Asymmetry Thesis that perduring objects do *not* "live" in temporal-like worlds: they are too long to fit in there. The fact that their momentary parts do provides an indirect sense in which one could speak of the coexistence relations among perduring wholes "taken" at certain moments of their individual times. But such relations are of secondary importance to the perdurantist ontology. Their failure to obey a "reasonable calculus" is, therefore, metaphysically inconsequential.

I submit that the best way to categorize such derivative relations is to treat them as *perspectival* relations in spacetime. A spatial analogy may be helpful here. A typical art museum has numerous exposition rooms variously connected to each other by wide entrances on their sides. There may be a location in one of the rooms from which one can see through several such openings and thus "line up" several rooms in a single line of sight. One can even manage to see some paintings or parts of them in distant rooms. Slightly changing one's location may significantly affect the arrangement of the rooms in the line of sight: some openings may disappear from it and others emerge. There may be a perspective in which Room 101, say, is lined up in this way with Rooms 102 and 104; another perspective in which Rooms 101, 103, and 104 are so lined up; and another in which Rooms 102, 103, and 104 are. But there may be no single perspective in which all four rooms could be lined up. This tells us something informative about the floor plan of the museum. But it has no further implications.

Similarly, the distribution of perduring objects in spacetime tells us something useful about the way their stages relate to each other. But given the central thesis of perdurantism—that *ordinary objects* are four-dimensional entities and not stages—we should view the spatiotemporal relations among the stages as merely perspectival features of a collection of perduring wholes, exhibited in the single spacetime manifold, not as fundamental facts about their existence and coexistence. Unlike enduring objects (and unlike stages), such wholes do not live in (and hence, cannot share) temporal-like worlds. What counts as a world for the former is no more than a perspective for the latter.

Contextuality is only part of the price endurantism has to pay in the transition to relativity. Besides being contextual, coexistence in the relativistic world of enduring and exduring objects may exhibit chronological incoherence.

7.8. Chronological Incoherence

Let us start with an unproblematic case. The 30-year-old Data coexists with the 35-year-old Captain Kirk and the 40-year-old Captain Picard: they share a common temporal-like world occupying HPS_1. As Data grows older and reaches the age of 35, he happens to coexist with the younger Kirk and Picard, who both have just turned 32: the three characters share a common temporallike world occupying HPS_2 (see Figure 7.7).

This, however, should not be viewed as being particularly disturbing. At any moment in his life, Data belongs to an infinite number of temporal-like worlds. The fact that one can pick out a chronologically incoherent series of such worlds (e.g. those occupying HPS_1 and HPS_2, in this order) should not be held against the endurantist or the exdurantist, as long as another, chronologically coherent series is available; for example, the series including HPS_1 and HPS_3. The temporal-like world occupying HPS_3 features the 35-year-old Data and the correspondingly older Kirk and Picard. One is not saddled with the offending sequence of HPS_1 and HPS_2, because there

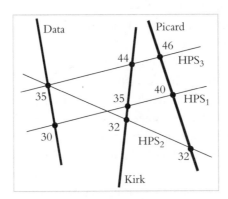

Figure 7.7. Relativistic Chronology.

is no reason to allow one to exploit the latitude inherent in relativistic spacetime frivolously, by sequencing temporal-like worlds at will. The availability of chronologically coherent series of such worlds is all that the *Star Trek* biographer needs to tell a sensible story about the life careers of the three famous characters and their relations to each other.

But there are cases where a chronologically coherent series of temporal-like worlds is not available (unless one makes such a series improperly short) and there is no escape from a disturbing conclusion that ageing results in becoming a contemporary of progressively younger companions. One case of this sort is sketched in Figure 7.8. Here the most one can do is identify a chronologically coherent series of temporal-like worlds containing the correspondingly ageing Data, Kirk, and Picard, for example, some series including HPS_1 and HPS_2, in this order. Adding Trevor to the picture, however, turns the series into a bad one. As Data, Kirk, and Picard all grow older, they find themselves in worlds with the younger and younger Trevor. And that is not the worst possible scenario yet. With some modifications, one could make progressively ageing Data, Kirk, and Picard unavoidable contemporaries of, first, Chief Trevor, next the 15-year-old Cadet Trevor, then the newly born Klingon baby just named Trevor, and eventually, Trevor's great-grand-grandfather! This is surely an unwelcome result. Along with contextuality, it brings out the difficulties of formulating multigrade coexistence relations in the relativistic world of enduring or exduring objects.[17]

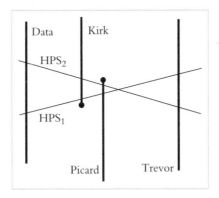

Figure 7.8. Chronological Incoherence.

[17] Cody Gilmore and Dean Zimmerman (in correspondence), and Ian Gibson and Oliver Pooley (in print: Gibson and Pooley 2006: 176 and n. 32) have all suggested that the proper lesson to draw from

Why is perdurantism not afflicted with the same or similar problems? One reason has been noted earlier: there is no strict sense in which perduring objects can be said to age. The bearing of considerations having to do with ageing on perdurantism is, therefore, indirect and metaphysically harmless. This is sufficient to uphold the asymmetry between perdurantism and both its rivals in the face of chronologically incoherent series.

But it is worth repeating that there is another and more general aspect of the asymmetry that pertains both to contextuality and chronological incoherence, even though the former is a "static" phenomenon, having to do with a collection of objects taken at particular moments of their individual times or in complete abstractions from them, whereas the latter is a "dynamic" phenomenon having to do with ageing. The feature common to both of them is that perduring objects do not occupy the sort of habitats (i.e. temporal-like worlds confined to single HPSs) that may exhibit contextuality or chronological incoherence of *coexistence*. All perduring entities coexist with each other in the Minkowski world. This does not mean that one cannot pose temporally qualified questions about the coexistence of their various parts. And the answers to them are informative as they tell one something about the overall distribution of perduring objects in spacetime. But the CASH value of such information is merely perspectival, not ontological.

Based on considerations of the last two sections I conclude that perdurance remains a preferred mode of persistence in the context of special relativity.

7.9. Some Objections

But this conclusion turns on attaching a lot of metaphysical weight to the notions of existence@ and coexistence in temporal-like worlds of Minkowski spacetime. One consideration in support of this strategy was derived from the modal–temporal analogy (§7.1): temporal-like worlds are,

the scenario is that any coherent story that includes all four characters must start with some of them not yet existing. I agree that such a story is available. My point, however, is that any story that begins with all four characters must have a *coherent continuation*. The scenario shows that this expectation may be frustrated.

in many ways, similar to the worlds of modal realism. But the analogy has its limits and cannot, by itself, convince those who may be inclined to regard coexistence in spacetime as a merely locational notion that is devoid of metaphysical significance. If that is the case then endurantists and exdurantists have no more to fear from the contextuality and chronological incoherence of coexistence than perdurantists do, the point made very clearly by Gibson and Pooley (2006: 174–7).

Their point was made, however, in the context of rejecting the assumption that gave the metaphysically serious treatment of coexistence and existence@ considerable plausibility. The assumption, which finds its direct expression in CASH, is that temporal-like worlds in special relativistic spacetime must be *flat*. In other words:

(FLAT*) Entities that stand in a multigrade relation of coexistence in Minkowski spacetime must be located (or q-located) on a single time hyperplane.

If flat spacelike hypersurfaces are given such a privileged status in regimenting coexistence (and therefore, existence@) then there is a prima facie reason to attach metaphysical distinction to them and their contents may deserve the title of *worlds* (namely, temporal-like worlds). On the other hand, if all spacelike hypersurfaces in Minkowski spacetime, flat and non-flat alike, are treated on a par—the approach favored by Gibson and Pooley—then the distinction loses ground and the best strategy may then be to abandon world-talk altogether.

Quite independently of these general considerations, if one allowed generic spacelike hypersurfaces of variable curvature in Minkowski spacetime to replace flat ones in the statements of various principles of coexistence, then contextuality and chronological incoherence would evaporate, because the offending situations described in the last two sections would not arise.[18] This is due to the following mathematical fact about Minkowski spacetime:

(FACT) For any set of pairwise spacelike separated points p_1, $p_2, \ldots p_n$ in Minkowski spacetime, there is a continuous three-dimensional spacelike surface containing $p_1, p_2, \ldots p_n$.

[18] For a more detailed discussion of the consequences of such replacement see Balashov 2005c: 34–6. See also Gibson and Pooley 2006: 175.

FACT is non-trivial, and the proof, on behalf of the objector, is given in Appendix 7.1 to this chapter.[19]

What can be said in defense of FLAT*? As one might expect, considerations in its favor are similar to those adduced earlier (§5.2) in defense of FLAT:[20]

> (FLAT)   In the context of discussing persistence in Minkowski space-time it is appropriate to restrict the locations and q-locations of persisting objects and their parts to flat achronal regions representing (subregions of) of moments of time in inertial reference frames.

The reasons against allowing non-flat locations (and q-locations) on a par with flat ones were, briefly, as follows: flat achronal hypersurfaces are (i) available in Minkowski spacetime, (ii) widely used in physics, and (iii) are needed to extend the notions of moment of time and momentary location (in a given reference frame) to the special relativistic framework. Depriving flat hypersurfaces of this privilege over arbitrary achronal hypersurfaces vis-à-vis issues of persistence would also jeopardize the related notions of momentary shape, momentary achronal composition, and the like, which would lose much of their ground and get completely out of touch with any familiar concepts.[21] Since special relativity does not *force* us to abandon them (i.e., since special relativity preserves the distinction between flat and non-flat achronal hypersurfaces) we are better off keeping them in place. Adherence to the following maxim is, therefore, desirable in doing physics-informed metaphysics:[22]

> (MOOR)   In adapting a metaphysical doctrine to a physical theory one should seek to minimize the degree of the overall ontological revision.

---

[19] To see that FACT is indeed non-trivial, note that a hyperplane defined by a triangle of three pairwise spacelike separated points in 2+1 spacetime need not be everywhere spacelike. Assume $c = 1$ and the following Cartesian coordinates of four points: $p_1(-10,0,0)$, $p_2(0,-10,11)$, $p_3(0,10,11)$, and $p(0,0,11)$. While p_1, p_2, and p_3 are pairwise spacelike separated, p_1 and p (which is the midpoint of the line segment p_2p_3) are not.

[20] In response to Gibson and Pooley's related objections, 2006: 160–5.

[21] We know what shape at a time is, even if time is relative to an inertial frame. But do we have any idea of shape at an arbitrarily curved achronal region?

[22] MOOR = Minimizing Overall Ontological Revision (see §5.2).

The same sorts of considerations apply to FLAT*. While FLAT concerns q-location, as well as momentary properties and composition of single spatially extended persisting objects or their parts, FLAT* concerns (non-trivial) coexistence of collections of objects. If the latter notion is to serve meaningful purpose in Minkowski spacetime, it must be similarly protected against diluting the distinction between appropriate and inappropriate geometrical underpinnings for it. In fact, the connection between FLAT and FLAT* is quite intimate for, arguably, coexistence of a collection of smaller objects is grounded in the same facts as composition (at a time in a frame) of a larger object constituted by that collection.

If FLAT* is supported by these considerations then temporal-like worlds in Minkowski spacetime emerge as flesh-and-blood entities against the background of various ghostlike curved structures. I believe this distinction and privilege are substantial enough to bear the weight of the arguments of §§7.7 and 7.8.

Before leaving the topic we must consider another worry raised by Gibson and Pooley (2006: 172), concerning the idealization of persisting objects as pointlike—an idealization adopted throughout our discussion of coexistence in Minkowski spacetime. This simplifying assumption was made to reduce achronal regions of interest—those that serve as locations and q-locations of persisting objects or their diachronic parts and underlie various relations of coexistence—to single spacetime points and to facilitate talk of the objects' proper time. For a pointlike object, proper time is simply the time measured along the one-dimensional path of the object. But the notion of proper time for spatially composite objects—those that have figured prominently in our discussion (i.e., Descartes, Captain Kirk, and so forth)—is far from being well defined. If so, then is the idealization as harmless as it appears?

There are two separate questions here: (i) Can the notion of proper time for a spatially extended composite object be defined? And even if not, (ii) does this make the decision to treat persisting objects as unextended point particles harmful for the arguments of this chapter? Both questions are generally important for the issue of persistence in Minkowski spacetime, and I shall consider them in turn.

(i) How could one go about defining the notion of proper time for a composite object whose parts may undergo complicated relative motion?[23] One obvious thought is to associate the proper time of the object in question with the time measured along the trajectory of its *center of mass*. But it is far from clear that the notion of center of mass is well defined in special relativity. In the classical case, one calculates the radius vector of the center of mass r_0 of a system of particles at a given moment of time by taking the weighted sum of the radius vectors of its components:

$$r_0 = \Sigma m_i r_i / \Sigma m_i$$

In special relativity, "a given moment of time" must be relativized to a frame. But *which* one? Presumably, to the object's *instantaneous rest frame*. But to know in which frame the object is "instantaneously at rest," in the general case of n particle components in complex state of relative motion, one apparently needs to know what trajectory in spacetime represents the motion of the "object as a whole," and it is unclear that this could be known without knowing the trajectory of the object's center of mass. We are therefore in a circle.

In fact, one can show that the instantaneous rest frame of a composite object is not a well-defined notion independently of evaluating the prospects of any candidate for the role of the center of mass. Following Gibson and Pooley (2006: 194, n. 29), consider a case of an object composed of two oscillating point particles of equal mass, moving uniformly towards and away from each other at the same speed (see Figure 7.9) in frame (x,t). The object as a whole is at rest in this frame (e.g., at t_1, t_2, t_3, etc.). But it is *also* periodically at rest in frame (x',t') co-moving with one of the particles (e.g., at t_1', t_2', etc.). Thus the object is at rest in *both* frames that are in relative motion!

This, in turn, seems to imply that there is no unique trajectory of the center of mass of this system. Indeed, one indisputable candidate for such a trajectory is the symmetry line of Figure 7.9, with respect to which the motion of both objects is perfectly symmetrical in frame (x,t). But given that the system is also periodically at rest in frame (x',t'), a line that includes stretches parallel to the common direction of the worldlines of both objects during their common periods of rest in (x',t') (and hence,

[23] I am grateful to Oliver Pooley, Nick Huggett, and John Norton for their help with this complicated topic.

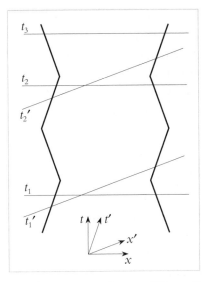

Figure 7.9. A composite object is at rest in two different frames of reference.

at an angle to the symmetry line) would *also* be a good candidate for a trajectory of the center of mass.[24]

This raises an interesting question: Is there *any* general way to define a trajectory representing, somehow or other, the motion of an arbitrary composite object in Minkowski spacetime? To fix ideas, assume that object o is composed of n particles $o_1, o_2, \ldots o_n$ with continuous and smooth trajectories $\mathbf{r}_i = \mathbf{r}_i(s)$, $t = t(s)$ in a coordinate system $(\mathbf{r}, t)$ adapted to some inertial reference frame (it does not matter which one), where s is a real-valued parameter. We are looking for a trajectory $\mathbf{r}_o = \mathbf{r}_o(s)$, $t_o = t_o(s)$ representing (somehow or other) the motion of o. Choose some particle, say, o_1 (it doesn't matter which one). For any s, begin by identifying a time hyperplane through $(\mathbf{r}_1(s), t(s))$, at which the total 3-momentum of o is zero. That is to say, identify a reference frame $F(s)$ (an "instantaneous rest frame of o"), such that, for some coordinate system $(\mathbf{r}^F, t^F)$ adapted to F, a particular time hyperplane $t^F = t^F(s)$ contains $(\mathbf{r}_1(s), t(s))$ and $|\Sigma m_i^F \mathbf{v}_i^F| = 0$,

---

[24] In correspondence, Cody Gilmore has suggested another interesting case illustrating the same point: an object composed of a linear array of *infinitely* many identical point particles, each receding from its neighbor at the same relative velocity. The spacetime trajectory of *any* such particle could then be taken to represent, equally well, the trajectory of the whole object—an extreme case in point! Below I abstract from such examples, which involve an infinite number of material parts, and focus on a system of n particles.

where all the $m_i{}^F \mathbf{v}_i{}^F$'s are calculated at $t^F = t^F(s)$ in $(\mathbf{r}^F, t^F)$. Then find the radius vector of the center of mass $C^F(s)$ of o at $t^F = t^F(s)$ in $(\mathbf{r}^F, t^F)$: $\mathbf{r}^F_o = \Sigma m_i{}^F \mathbf{r}_i{}^F / \Sigma m_i{}^F$. Finally, obtain the position $(\mathbf{r}_o(s), t_o(s))$ of $C^F(s)$ in the original coordinate system $(\mathbf{r}, t)$.

Questions remain: Does the procedure generate a unique, continuous and smooth trajectory? The case depicted in Figure 7.9 raises doubts at least about uniqueness. But perhaps the "latitude" inherent in such cases is due to the small number of particles involved ($n=2$) and disappears when n becomes sufficiently large? I do not have answers to such questions.[25]

(ii) But even in the absence of general answers to such questions, the impact of the problem on the arguments of this chapter is minimal. Suppose the objects involved in an interesting relation of coexistence are composite and spatially extended. The basic fact underlying their coexistence remains the same: the presence of a single HPS containing the locations or q-locations of all the objects or their diachronic parts. Of course, each such (q-)location is now a set of points (i.e., a scattered 3D region comprising the (q-)locations of the object's constituent particles, if objects endure or exdure, or of their diachronic parts, if objects perdure), rather than a single point.

What is different is that the exact determination of the *age* of the enduring or exduring objects involved in the relevant coexistence relation becomes difficult, if not impossible. This can be done only approximately, with a certain "margin or error." The main factor contributing to the "fuzziness" of an object's age is the relative motion of its constituent particles, whereby the ages of different particles get progressively "out of step" with each other, due to relativistic time dilation. A reasonable estimate of the scale of this effect can be obtained for different kinds of objects by taking into account the factor corresponding to the characteristic speed of the relative motion of the object's parts. Thus for human beings the relevant speed can be associated with molecular motion, with a conservative upper bound set at 1 km/sec. This corresponds to $\gamma = 1.000000000006$ and translates into the cumulative time difference (between the "ages" of two molecules in

---

[25] Extreme "latitude" of this sort is also present in Gilmore's case (see the previous note) of infinitely many point particles. The procedure sketched above appears to be similar to one of the proposals discussed in a more technical environment by Pryce 1948. See also Schattner 1978 for a more recent discussion of the notion of center of mass in general relativity. I thank Oliver Pooley for pointing me to these valuable sources.

constant relative motion) of a mere 0.01 sec over the period of 50 years. Clearly this sort of "fuzziness" is completely innocuous and can be safely ignored in statements about coexistence involved in the arguments of this chapter.[26]

Appendix 7.1:   Proof of FACT

Consider a set of points p_1, p_2, ... p_n in Minkowski spacetime. For all $i \neq j$, p_i and p_j are spacelike separated. Prove that there is a continuous spacelike 3D surface[27] containing p_1, p_2, ... p_n.

> Def.: A function f on a metric space X is K-Lipschitzian just in case $\exists K > 0$ such that for all $x, x' \in X$, $|f(x) - f(x')| \leq Kd(x, x')$.

> Lemma: If functions f_1, ... f_n on X are K-Lipschitzian, then their lower bound $f \equiv \inf_{i=1...n} f_i$ is K-Lipschitzian.

Proof of Lemma, by induction:

1. *Base step.* Assume f_1 and f_2 are K-Lipschitzian. Fix two points x and y. Let $g \equiv \inf (f_1, f_2)$. We have:

$$-Kd(x, y) \leq f_1(x) - f_1(y) \leq Kd(x, y) \tag{1}$$

$$-Kd(x, y) \leq f_2(x) - f_2(y) \leq Kd(x, y) \tag{2}$$

Suppose $f_1(x) \leq f_2(x)$. (Otherwise exchange f_1 and f_2.) Then $g(x) = f_1(x)$. Consider two cases: (a) $f_1(y) \leq f_2(y)$ and (b) $f_1(y) > f_2(y)$.

(a) If $f_1(y) \leq f_2(y)$ then $g(y) = f_1(y)$ and, from (1): $-Kd(x,y) \leq g(x) - g(y) \leq Kd(x,y)$.

(b) If $f_1(y) > f_2(y)$ then $g(y) = f_2(y)$. From (1): $g(y) = f_2(y) < f_1(y) \leq f_1(x) + Kd(x,y) = g(x) + Kd(x,y)$.

Therefore, $g(y) - g(x) < Kd(x,y)$.

---

[26] That is, in speaking about the coexistence of the 50-year old Kirk and the 32-year old Data, etc.

[27] For our purposes, a surface is spacelike just in case any two points on it are spacelike separated; we do not require differentiability. But we shall accept, without proof, that if a sought-for continuous (but perhaps "corrugated") surface is available, it can always be appropriately "smoothened" and made everywhere differentiable, thus yielding, in a technical sense, a *hyper*surface.

From (2): $g(y) = f_2(y) \geq f_2(x) - Kd(x,y) \geq f_1(x) - Kd(x,y) = g(x) - Kd(x,y)$.

Therefore, $g(y) - g(x) \geq -Kd(x,y)$.

2. *Inductive step.* Assume: If functions $f_1, f_2, \ldots f_n$ on X are K-Lipschitzian, then $f \equiv \inf_{i=1\ldots n} f_i$ is K-Lipschitzian. Show: If functions $f_1, f_2, \ldots f_{n+1}$ on X are K-Lipschitzian, then $f \equiv \inf_{i=1\ldots n+1} f_i$ is K-Lipschitzian.

Denote: $g \equiv \inf_{i=1\ldots n} f_i$. By assumption, g is K-Lipschitzian. Note that $f_i = \inf(g, f_{n+1})$. Apply the result of the base step to g and f_{n+1}.

Main proof. Pick a frame. Set $c = 1$. We are looking for a continuous real function F on X satisfying $|F(x) - F(y)| < d(x,y)$ for all distinct x and y and such that $F(x_i) = t_i$, $i = 1, 2, \ldots n$.

$$\text{Let } K = \min_{\substack{1 \leq i \leq n \\ 1 \leq j \leq n \\ i \neq j}} \frac{|t_i - t_j|}{d(x_i, x_j)} \text{ and } F_i(x) \equiv t_i + Kd(x, x_i).$$

Obviously $K < 1$. Define $F \equiv \inf_{i=1\ldots n} F_i$.

By the triangle inequality,

$$|F_i(x) - F_i(y)| = |Kd(x, x_i) - Kd(y, y_i)| \leq Kd(x, y).$$

Therefore all $F_i(x)$ are K-Lipschitzian. By Lemma, $F(x)$ is K-Lipschitzian. Hence $F(x)$ is continuous and $|F(x) - F(y)| < Kd(x,y)$. Since $K < 1$, this entails, for all distinct x and y, $|F(x) - F(y)| < d(x,y)$. It remains to show that $F(x_i) = t_i$, $i = 1, 2, \ldots n$. First, note that $F_i(x_i) = t_i$. Next, for all $j \neq i$, $F_j(x_i) \geq t_i$. Hence $F(x_i) \equiv \inf_{j=1\ldots n} F_j(x_i) = t_i$.[28]

---

[28] The proof strategy and the choice of F are due to Tom Goodwillie. I was also greatly helped by Valery Alexeev who patiently educated me on Lipschitzian functions. I thank both of them.

8

Shapes and Other Arrangements in Minkowski Spacetime

The foregoing discussion adopted a rather course-grained view of the population of our world, which abstracted from the spatial dimensions of individual material objects. By contrast, considerations of this chapter put emphasis on spatial extension, which finds its manifestation in the objects' *shape* and the fine details of their mutual arrangement. The relativistic behavior of these properties is, in many ways, remarkable, as it is counterintuitive. The phenomenon of length contraction (see §3.6) lies at its basis. In its turn, length contraction is grounded in the fundamental facts about the relativity of simultaneity in Minkowski spacetime.

There is every reason to regard length contraction and other relativistic effects (such as time dilation and velocity addition) as *perspectival* phenomena in spacetime that are similar to more familiar perspectival phenomena in space. I believe relativistic perspectivalism favors a particular mode of persistence, namely perdurance, over its major 3Dist rival, endurance. This chapter defends an explanatory argument to this effect.[1] The argument was first developed in Balashov 1999 and 2000c[2] and has since come under criticism.[3] Below I revisit the original argument and respond to my critics.[4]

---

[1] It is less clear whether the argument favors perdurance over *exdurance*. If not then the argument could be taken to support 4Dism against 3Dism. In any case, the argument is best stated in terms of the opposition of endurance and perdurance, and this is how I develop it below.

[2] The argument's pedigree goes back to Quine 1960: 172, 253 ff; 1987 and Smart 1972 who have sketched highly suggestive and heuristically valuable comments in a similar context, which, however, came short of constituting persuasive arguments. See §5.3 for a detailed discussion of Quine's and Smart's early views of persistence in relativistic spacetime.

[3] In the works of Sider 2001: 79–87; Miller 2004: 66–8; Gibson and Pooley 2006: 187–91; Sattig 2006: 182–3; Gilmore 2008: §5, and a referee's comments on an earlier draft of this chapter.

[4] I hasten to acknowledge that a portion of the criticism is well deserved: the way in which I put some points in Balashov 1999 and 2000c was unfortunate and misleading. The critics' reaction to those

I would like to begin, however, by presenting an especially striking case of the relativity of shape in Minkowski spacetime.

8.1. How Rigid is a Granite Block?

Consider a granite block moving with velocity v (which is a considerable fraction of the speed of light) and suspended from vertical threads that move along with it (see Figure 8.1).[5] At a certain moment all threads are cut and the block starts to fall, continuing at the same time its inertial horizontal motion. Figure 8.2 represents a series of snapshots showing the block at some moments in this process.[6] Figure 8.3 represents a similar series of snapshots taken in the original rest frame of the block.

The block remains straight in the first series but becomes progressively bent in the second. How could it be? There might, initially, be two worries about it. First, the block is made of granite and thus simply cannot bend. (If you think granite is insufficiently rigid, pretend that the block is made of *super*granite.) Second, the block cannot *both* remain straight *and* undergo bending.

These worries are, of course, misplaced. The block does both things. And it bends no matter how rigid its material is. Moreover, it always bends in the same way. How so? The key, as always, lies in the relativity of simultaneity. The threads suspending the block are cut *simultaneously* in the "laboratory frame" resulting in free fall of all segments of the block

Figure 8.1. Granite block in horizontal motion.

---

points was generally charitable; accordingly, their objections focus on more important issues—those considered below. Two related topics, explanatory virtues and explanatory relevance, loom especially large. I am indebted to my critics for pressing them.

[5] The essential details of the scenario come from Sartori 1996: 185–90, where it is used to illustrate one of the lesser-known "paradoxes" of special relativity, first introduced by Wolfgang Rindler 1961.

[6] Note that Figures 8.1–8.3 are *not* spacetime diagrams, but series of merely spatial "snapshots" taken at different moments of time in two reference frames.

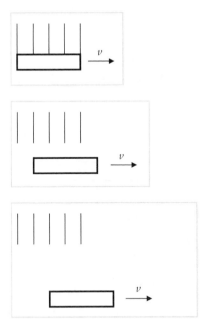

Figure 8.2. Granite block in free fall, continuing to move horizontally.

(see Figure 8.2). In the original rest frame of the block, however, the cutting events occur *successively* (seen in Figure 8.3). When the rightmost thread is cut the part of the block previously held by it begins to fall immediately. But the rest of the block remains horizontal. By the time the next thread is cut the segment of the block just underneath it still "does not know" that the rightmost part is already in free fall and, hence, does not have a chance to exert a sheer force that could stop the bending of the right end of the block. Why? Remember that the cutting events are simultaneous in the laboratory frame, hence, spacelike separated from each other. Therefore, no physical influence can propagate from one such event to the next. Nothing can stop a given segment of the block from free fall, once the thread holding it is cut. Accordingly, nothing can stop the block from bending. The strength of the material is beside the point.

Along with some other "paradoxes," this scenario is often taken to show that there are no rigid bodies in special relativity, that is to say, no bodies that can keep their shape invariant, even in the idealized limit.[7]

---

[7] See, e.g., Sartori 1996: 184–5.

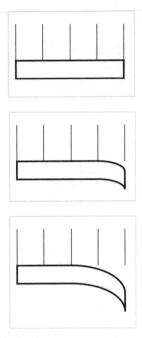

Figure 8.3. Granite block in "free fall" with snapshots taken in its rest frame.

(Thus supergranite is of no help.) Shape and other arrangements in 3D space are, in this theory, perspectival phenomena. But there must be something permanent standing behind all the different perspectives, such as those shown in Figures 8.2 and 8.3. What stands behind them is, of course, a 4D perduring object with an invariant 4D shape.[8] Or so I shall argue.

8.2. Perspectives in Space

The phenomenon of perspectivalism is familiar. It turns on an important connection between invariance and objectivity.[9] Any object can present itself differently in different perspectives. But there is something permanent (hence, objective) standing behind all such perspectival representations.

---

[8] I make no attempt to depict a 4D perduring granite block.
[9] See §3.2 and 3.4 for a brief discussion of this connection in the context of Galilean spacetime.

Perspectives in space are especially telling. Suppose you observe a collection of 2D shapes (Figure 8.4). Are they related? Not obviously, until you realize that these shapes are perspectival representations of a single object, a 3D cube, whose 3D shape is *invariant* (Figure 8.5).

Consider another example (Figure 8.6). Does the rotunda fit in between the trees? Speaking two-dimensionally, there is no fact of the matter, because both pictures are perspectival representations of the same 3D reality (see Figure 8.7).

In both cases, to explain the relation between the first series of pictures, one needs to invoke an extra dimension—the third dimension of space. This is what allows one to recognize the series as mere perspectival representations of an underlying reality.

We know by now that, just as there are perspectives in space, there are perspectives in spacetime.

Figure 8.4. Two-dimensional perspectival shapes.

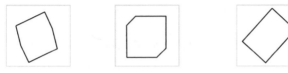

Figure 8.5. Three-dimensional cube in different perspectives.

(a)   (b)

Figure 8.6. Rotunda and trees.

Figure 8.7. Rotunda and trees in three dimensions.

8.3. Perspectives in Spacetime

Suppose you observe a collection of 3D shapes (Figure 8.8):

Figure 8.8. Collection of three-dimensional shapes.

Are they related? Not obviously, until you realize that these shapes are perspectival representations of a single object whose shape is invariant. But what *sort* of thing must the object be, in order to present itself in such different ways in various perspectives, without being different from itself? The answer is easily anticipated: the object must be *four*-dimensional (4D); it must be extended in time as well as space. It will then have different 3D shapes in different perspectives (associated with different inertial reference frames and related by Lorentz transformations), because such shapes will be intrinsic properties of its 3D parts. What stands behind, and thus explains, the whole variety of 3D shapes is a single 4D entity. It is not so easy to depict it; but Figure 8.9 gives an idea.

This explanation is open to the perdurantist, who believes in 4D objects, but not to the endurantist, who denies their existence. Indeed, the endurantist will have a hard time explaining how "separate and loose" 3D shapes come together in a remarkable unity, by lending themselves to an arrangement in a smooth 4D volume. Where the perdurantist has a ready and natural explanation of this fact: different 3D shapes are cross-sections of

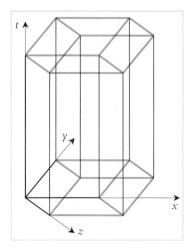

Figure 8.9. Four-dimensional cube in spacetime.

a single 4D object, the endurantist must regard it as a mystery. One should not expect to be able to fit an arbitrary collection of 3D shapes into a neat 4D shape, without corrugations, dents, and gaps.[10]

As another example, consider a 10-meter pole moving at the speed $v = 2.6 \times 10^8$ m/sec through the open doors of a 10-meter barn.[11] Will the pole fit completely inside the barn? First, consider the situation from the point of view of the barn. In the barn rest frame, the pole is Lorentz-contracted to 5 meters and thus perfectly fits in (see Figure 8.10(a)). In the pole rest

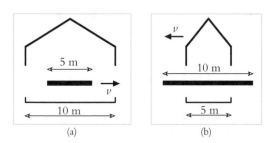

Figure 8.10. The pole and barn scenario: (a) in the barn's rest frame; (b) in the pole's rest frame.

[10] This can be seen easily from a spatial analogy. Replace one of the hexagons in Figure 8.4 with a heptagon and *then* try to fit the resulting 2D shapes into a neat 3D volume.
[11] The pole and barn paradox considered below figures prominently in relativity textbooks. My exposition owes much to Sartori 1996: §6.2.

frame, on the other hand, it is the barn that is Lorentz–contracted (and becomes only 5 meters wide).

Consequently, both ends of the pole will protrude from the barn for a short time interval and the pole will never fit entirely inside the barn (see Figure 8.10(b)). So will the pole fit inside or won't it?

To make the situation a bit more dramatic, suppose the rear door of the barn is initially shut and the front door open. As soon as the trailing end of the pole has cleared the front door, the door gets shut. And when the forward end of the pole is about to collide with the rear door, it is opened thus letting the pole pass through. From the barn point of view, at a certain point, the pole finds itself completely inside the barn, both doors shut (see Figure 8.11). From the pole point of view, the pole is never completely inside but it also never collides with any door: for a short interval, the pole protrudes from both open doors of the barn (see Figure 8.12).

So will the pole fit inside the barn or will it not? The paradox is resolved by taking into account the relativity of the temporal order of two events: E_1, at which the forward end of the pole passes through the rear door (while the door opens immediately before), and E_2, at which the trailing end of the pole passes through the front door (and the door is shut immediately

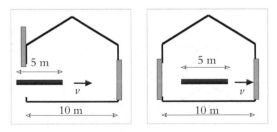

Figure 8.11. The pole is completely inside the barn, with both doors shut.

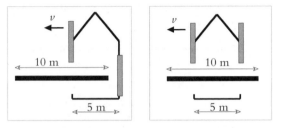

Figure 8.12. The pole protrudes from the open doors of the barn.

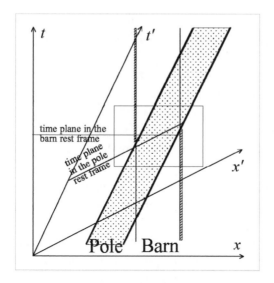

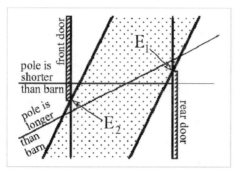

Figure 8.13. The pole is shorter than the barn in the barn rest frame and longer than it in the pole rest frame.

thereafter). In the rest frame of the barn (x,t), E_1 occurs after E_2 thus allowing the pole to be entirely inside the barn. In the pole frame (x',t'), E_1 happens before E_2 thus allowing the pole to protrude from both ends of the barn (shown in Figure 8.13).

So in the end, both parties are right: there is simply no fact of the matter as to what event happens first and, thus, no conflict between two relative, perspective-restricted versions of the scenario.

Both the endurantist and the perdurantist may be required to provide a metaphysical account of what goes on in the pole and barn scenario. The perdurantism story is rather straightforward. One should simply take the spacetime diagram in Figure 8.13 literally. It features two four-dimensional objects: the pole (occupying the shaded area) and the barn, and various relations among their temporal parts. Thus the temporal part of the pole at a certain moment of time in the barn frame is shorter than the corresponding temporal part of the barn, whereas the temporal part of the pole at a particular moment of time in its rest frame is longer than the corresponding temporal part of the barn. These seemingly contradictory relations hold, not because the objects change their length (the scenario involves no such change), but because the pole and barn are "viewed," in their mutual relationship, from different angles in spacetime (just as the rotunda and the trees are viewed from different angles in space in Figures 8.6(a) and 8.6(b)). But in order for them to be capable of being viewed in so different perspectival ways, both objects must be four-dimensional. Their invariant 4D configuration stands behind apparently conflicting frame-restricted 3D representations. Four-dimensionality is invoked to explain the conflict away (just as the third dimension of space is invoked to explain away the seeming conflict between the two pictures of the rotunda and the trees).

This is not to suggest that the endurantist has *nothing* to say, by way of explaining the different 3D shapes of the object represented in Figure 8.8 and of the mutual arrangement of the 3D pole and the 3D barn represented in Figures 8.11 and 8.12. The point rather is that the endurantist explanation does not go far enough and is, for that reason, inferior to the perdurantist account. In the case depicted in Figure 8.8, the endurantist will invoke a multitude of 3D shapes indexed, one way or another, to different moments of time in different inertial frames.[12] In the pole and barn case, what needs to be so indexed is the mutual arrangement of two 3D objects in space (in the simplest case, in terms of the relation *being longer than*). But after this is said and done, the question remains: *why* are all the perspective-indexed 3D shapes and arrangements so nicely related in *four* dimensions, as shown in Figures 8.9 and 8.13? The best explanation, to repeat, is to take these figures literally as representing the objects' extension in four dimensions.[13]

---

[12] See the end of §5.1 for alternative approaches to such "indexing," on behalf of the endurantist (and the exdurantist).

[13] In this respect Gilmore, 2008: 1234–5, may be right to point out that the argument directly supports the thesis that objects have temporal extent and only indirectly the thesis that objects

The above explanatory argument from special relativity to perdurantism has come under criticism. In the remainder of this chapter I respond to my critics. To keep the discussion reasonably concise I focus on the case of a single object (Figures 8.8 and 8.9) and set the pole and barn scenario aside.

8.4. Are Shapes Intrinsic to Objects?

Let us go back to the point just made: in order to explain the unity behind the 3D shapes one must view them as perspectival representations of a single 4D shape. The perdurantist, I have argued, is best positioned to offer such an explanation; on her theory, the 3D shapes of 3D material entities (that is, the temporal parts of a perduring object) result from "viewing" the 4D shape of a single 4D material entity from various "angles" in spacetime. There is a very natural functional relationship here between the region occupied by the entire perduring object and the regions occupied by its temporal parts. The latter regions add up to the former in the same way in which the material parts add up to the material whole.

One feature that seems to prevent the endurantist from matching this achievement is a categorical difference between the nature of 3D shapes of an enduring object at times-in-frames and the nature of the 4D shape of the enduring object's path in spacetime. The 3D shapes (or the having of them) are irreducibly indexed to a frame-relative time, but the 4D shape is not. Given these different categories of shapes, it is hard to see how the former can "add up" to the latter. The 4D shape is absolute (i.e. had *simpliciter*) in a way the 3D shapes are not, in the ontology of endurance.

have temporal parts. In my approach (§§2.4, 4.1, and 5.1) both theses are part and parcel of perdurance and the denial of both is part and parcel of endurance. But as noted earlier (§2.1), some philosophers disagree and draw the line differently. See, in particular, Hawthorne 2006; Parsons 2007; and Saucedo, forthcoming. In Gilmore's own approach (2006: §3 and 2008: §4), the two issues—that of having temporal extent and that of having temporal parts—are kept apart, resulting in a fourfold classification of the modes of persistence. See §2.5 for a discussion of Gilmore's classification. The need for "unnatural" combinations (e.g. "endurance" with temporal parts or "perdurance" without temporal parts) almost always comes from a desire to accommodate exotic possibilities (e.g. instantaneous statues or extended simples), from which the present study abstracts. See §§1.2 and 2.4.6.

But there is a version of endurantism that appears to restore parity with perdurantism on *that* score.[14] The endurantist can deny that objects have irreducibly indexed 3D shapes. Rather, they have *absolute* 3D shapes *derivatively*, in virtue of being multilocated at appropriately shaped 3D regions, while lacking intrinsic shapes *themselves* altogether. Absolute (namely, not indexed) 3D shapes of such regions can "add up" to an absolute 4D shape of a 4D region (i.e. the object's path) just as naturally as they do in the ontology of perdurance (i.e., without cutting across categorical distinctions), thereby allowing the endurantist unhindered access to the explanatory resources of perspectivalism. But instead of dealing with the shapes of material entities we are now working exclusively with the shapes of spacetime regions and their subregions, which, if anything, simplifies the explanatory task.

To the advantage of this account, it accords well with an independently motivated view, recently defended by Thomas Sattig (2006), that the primary bearers of *all* temporary properties of enduring objects—3D shapes, as well as temperature, color, hunger, and so on—are achronal spacetime regions at which such objects are multilocated. But questions arise. If enduring objects do not have shapes intrinsically, what should we say about perduring objects and their temporal parts? It would seem that the view denying the intrinsic nature of shapes is fundamental enough to cut deeper than the endurance–perdurance distinction. This would impose on the perdurantist the notion that no material entities of his ontology (i.e. neither total perduring objects nor their temporal parts) have shapes by themselves. In the case of endurantism, this strategy could be motivated by concerns about multilocation and the problem of temporary intrinsics. But no such concerns are present in the ontology of perdurance.

Relatedly, what is left of the basic idea of *exact location*? On the view adopted throughout this book (see §2.2) and shared by most participants in the persistence debate, a spacetime region at which an object is exactly located is a region into which the object exactly fits and which has exactly the same size, shape, and dimensionality as the object itself. But on the proposal currently under consideration, the object itself has none

---

[14] I thank a referee for impressing upon me the importance of addressing this version of endurantism.

of these properties, so the relation of "fitting" is simply inappropriate. And without this relation, the concept of exact location becomes rather problematic.

It might perhaps be argued that these considerations simply beg the whole question of rejecting the prevailing doctrine that material objects have shapes intrinsically rather than merely derivatively. The derivative having of shapes would arise from the objects' being exactly located (whatever that means) at appropriately shaped regions. For all we know, the strategy might be consistently developed and lead to important insights.[15] Be that as it may, we need to set it aside here and consider other objections against the explanatory argument of §8.3 that were posed in the framework of the dominant view of shapes.

8.5. The Causal Objection

The moral of the spatial analogy with the cube (see Figures 8.4 and 8.5) was this: to provide a unified explanation of a variety of 2D shapes one has to upgrade the number of dimensions possessed by objects to three. Once this is done, the 2D states of affairs are easily recognized for what they are—mere perspectival representations of the underlying 3D reality. The same sort of consideration drives the relativistic argument. To provide a unified explanation of a variety of 3D shapes the number of the object's dimensions must be upgraded to four, whereby the 3D states of affairs can be recognized as mere perspectival representations of an underlying 4D reality. The difference, however, is that the dimension added in the second case is that of time. But the latter is widely regarded as a dimension of *causation*, and this raises the question of whether, instead of attributing the fourth dimension to objects, one could not simply point out that various 3D shapes of enduring objects result from the objects' causal evolution in time, in accordance with the laws of nature. In other words, what unifies the shapes is not the existence of a 4D entity, which is sliced up in different spacelike directions, but an underlying three-dimensional dynamics governed by the familiar physical principles. Taking this into consideration would enable

---

[15] See McDaniel 2003 and Skow 2007 for recent discussions.

the endurantist to match the explanatory achievements of the perdurantist. Kristie Miller notes:[16]

[I]f all we had were relativistic three-dimensional shapes and no theory about how they "fit together," we would be surprised to discover that they fill the volumes that they do. The theory of special relativity, however, along with various other laws of nature, allows us to predict how objects that exist in the present, will exist in the future. That is, they allow us to predict what the four-dimensional volume of an object will be.   (Miller 2004: 368)

But in this form, the causal objection misfires. The sequence of shapes represented in Figure 8.8 is *not* a causal sequence. It was not intended to describe an evolutionary process whose earlier stages determine, in accordance with the laws of nature, its later stages, in a particular *single* frame of reference. Rather, the configurations depicted in Figure 8.8 represent 3D shapes of the object at different moments of time in *different* frames of reference. Speaking geometrically, they feature the shapes of various spacelike cross-sections of the worldtube of the object, which are drawn randomly at different "angles" in spacetime. For all we know, some such configurations may correspond to *crisscrossing* hyperplanes of simultaneity, whose *overall* contents, arguably, cannot be causally related.

The case is easily illustrated (Figure 8.14).[17] Consider two crisscrossing spacelike "slices" R_1 and R_2 through a region of spacetime R. Suppose R_1 and R_2 are occupied by enduring objects o_1 and o_2 respectively and, to avoid begging any questions, leave it open whether $o_1 = o_2$. One can argue (see Gilmore 2006: 214 ff) that the *overall* contents of R_1 and R_2 cannot be related as cause and effect, for, intuitively, a sufficiently large portion of the contents of R_2 (that is, $R_2{}^f$) is to the future of a sufficiently large portion of the contents of R_1 ($R_1{}^p$), whereas another sufficiently large portion of the contents of R_2 ($R_2{}^p$) is to the past of the corresponding sufficiently large portion of the contents of R_1 ($R_1{}^f$). Consequently, the total content of R_1 cannot stand in a causal relation to the total content of R_2.

---

[16] The causal issue was first raised in the present context by Hud Hudson; see Balashov 1999: 660, n. 3.

[17] We have already encountered this case before (see §5.6) in discussing Gilmore's objections to crisscrossing slices. Figure 8.14 is adapted from Gilmore 2006: 215.

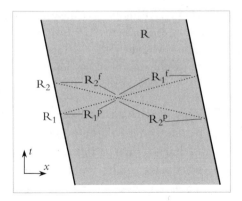

Figure 8.14. Crisscrossing slices.

This does not prevent the endurantist from telling a plausible causal story about the relation among the 3D shapes. But the story must be more involved. It must include a *mereological* component, along with the causal one. It is clear that, even though the overall contents of R_1 and R_2 (in Figure 8.14) are not related as cause and effect, their *piecemeal* contents are. As we know, a collection of 3D regions of Minkowski spacetime (such as R_1 and R_2, or the regions represented in Figure 8.8) may be multiply occupied by a single enduring object, whose pointlike parts may have causally cemented individual careers. This suggests a better strategy for restoring explanatory parity with perdurantism.

8.6. The Micro-Reductive Objection

Theodore Sider notes, on behalf of the endurantist:[18]

To have a shape at a time is just to have *parts* that are spatially related in a certain way at that time. . . . [R]elative to a chosen reference frame we can account for spatial relations between fundamental particles at times. Provided the three-dimensionalist can make sense of the part-whole relation in a relativistic context, then, she can account for the shapes of macroscopic objects in various reference frames. All of the perspective-indexed shapes are the result of a single set of facts about the

[18] As far as I can see, Thomas Sattig's critical comments, 2006: 182–3, could be read in the same spirit, even by someone who does not share his thesis that the primary bearers of temporary properties of enduring objects are achronal spacetime regions at which such objects are multilocated.

enduring object, which include (1) the holding of the part-whole relations, and (2) the holding of the occupation relation between fundamental particles and points of spacetime. This also explains why the shapes fit into the 4D volume that they do. The volume is generated as the sum of all the points occupied by the parts of the object; the shapes are slices of this volume.   (Sider 2001: 83)

The idea is this. Starting with the physical facts about multilocation of fundamental enduring particles at spacetime points, the endurantist could put her finger on the worldlines of such particles, to find out what spacetime points are occupied by what particle. She could note, next, that a given 3D object at a given time in a given reference frame is composed of a definite collection of fundamental particles. Finally, she could use this information to assemble together the worldlines of the fundamental constituents of a particular 3D object in spacetime. Such lines would fill a nice 4D volume, thereby resolving the "puzzle" about 3D shapes.

In Balashov (1999), I argued that while Sider's account provides the necessary elements of *some* explanation of the relation among the different 3D shapes that a single enduring object exhibits in different reference frames, such a "micro-reductive" account is explanatorily *deficient*.[19] The perdurantist has a much *better* explanation. The reason is that a good explanation must be *ampliative*: it must enhance understanding of a range of facts by invoking a *different* type of fact that would *unify* the former in a relevant way. No explanatory gain is achieved by merely restating the explanandum. But the latter is precisely what the endurantist "micro-reductive" strategy boils down to. Why does an enduring object, which exactly occupies a regular cubical region of space $R^{\blacksquare}$ at a certain time in a certain reference frame, also occupy a "skewed" region of space $R^{\blacklozenge}$ at some time in another frame (see Figure 8.8)? Sider's answer essentially is: because the object's particles have moved, in their different ways, from $R^{\blacksquare}$ to $R^{\blacklozenge}$ or vice versa. While this is certainly correct, it is not particularly enlightening. Contrast it with the perdurantist's account: $R^{\blacksquare}$, $R^{\blacklozenge}$ and many other peculiarly-shaped 3D regions are related because they are carved out from a smooth 4D spacetime region $R^{\square}$ occupied by a *single* object. The difference is quite similar to the

---

[19] At that time I was aware of Sider's then-unpublished objection (which later appeared in Sider 2001: 79–87).

difference between "explaining" various natural phenomena (e.g. chemical explosions, superconductivity, the increase of entropy in the universe) by "grounding" them in the totality of micro-physical facts on which they obviously supervene (e.g. the facts about the location of each of the micro-particles at every moment of time in an appropriate frame) and explaining them by invoking various mechanisms (certain types of chemical reaction, Cooper pair formation, "course graining") that unify phenomena and enhance their understanding.

The issue, in short, is not about proper *grounding*: we can agree that the facts about the occupation of spacetime points by fundamental particles, along with the facts about composition at-a-time in-a-frame, ground the facts about 3D shapes.[20] The issue is rather about *explanatory relevance* and *explanatory strength*. These features are crucial because the original argument is, in essence, an inference to the best explanation.

This point cannot be overemphasized. In the next section I put the point in yet a different perspective.

8.7. Pegs, Boards, and Shapes

In a number of related works Hilary Putnam discusses the shortcomings of micro-reductive explanations of various macrophenomena (including, eventually, mental life) by using the following example, which, I think, bears close resemblance to the issue of relativistic shapes (see Figure 8.15):

Suppose we have a very simple physical system—a board in which there are two holes, a circle one inch in diameter and a square one inch high, and a cubical peg one-sixteenth of an inch less than one inch high. We have the following very

[20] Ian Gibson and Oliver Pooley note rightly that this grounding does not go far enough and conclude that, for this reason, Sider's objection is incomplete. One should ask *why* a certain collection of spacetime points is occupied by a single fundamental particle. The answer, of course, comes from physics: "The various local fields around a particle determine where it 'next' is: such fields again determine where it is 'after' that; and so on until we have a complete worldline" (Gibson and Pooley 2006: 290). I fully agree that Sider's "micro-reductive explanation" can be further grounded in this way. And I do not see why Sider should not accept this as a friendly amendment. My point, however, remains: even with this additional grounding, the "micro-reductive explanation" is deficient. More on this below.

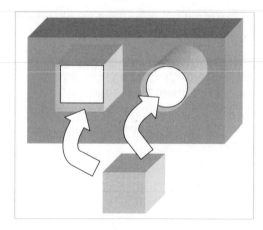

Figure 8.15. Board and peg.

simple fact to explain: *the peg passes through the square hole, and it does not pass through the round hole.*   (Putnam 1975: 295)

One could offer two different explanations of the fact. The first would include the physical details of the microstructure of the board and the peg, along with the laws of particle dynamics, which would show that, among all the possible trajectories of a complex physical system constituting the peg, there are some that would allow it to pass through another complex physical system constituting the section of the board with the square hole, but there are no such possible trajectories involving the section of the board with the round hole. The other explanation would simply cite the above noted geometrical properties of the objects.

Putnam argues that the "micro-reductive" explanation is vastly inferior and that everyone can see it. The superiority of the "macroscopic" explanation is based on the fact that it brings out the relevant structural features of the situation (i.e. the geometrical shapes of the peg and the holes) and abstracts from its irrelevant features (i.e. the inscrutable amount of physical detail and the complexity of the physical laws involved), thereby achieving genuine and unified understanding of the situation (see Putnam 1975: 296). There are, after all, other peg and board systems that have appropriately fitting shapes, but very few, if any, that have the same atomic composition. What needs explaining is not an unmanageable multitude of highly complex physical configurations and the resulting trajectories, but a quite manageable multitude of macro-facts about certain fitting and

SHAPES AND OTHER ARRANGEMENTS 215

non-fitting pairs of shapes. And all such facts are explained by invoking simple geometrical relations (plus the notion of rigidity).

I submit that the case of relativistic shapes is rather similar. The perdurantist has no need to invoke irrelevant micro-physical facts about the occupation of spacetime points by the fundamental constituents of material objects. What she is required to explain is not a multitude of disparate facts about the arrangeability of certain collections of 3D shapes in neat 4D volumes, but rather a general fact that some such collections all have the relevant "dispositional" property. And this general fact is explained by making an equally general point that such collections are cross-sections of 4D volumes occupied by real 4D material objects.

8.8. Perduring Objects Exist

But what, exactly, is involved in saying that 4D material objects are *real*? Gibson and Pooley note:

We agree that if objects perdure then the three-dimensional shapes are cross-sections through those four-dimensional objects. The question, though, is whether Balashov is entitled to simply *assume* the existence and shape of four-dimensional objects, only for this to then ground facts about the three-dimensional parts. Balashov thinks this is right and proper, claiming that "such parts are 'carved out' from a pre-existing ontological entity . . ." [2000: 333]. Yet there is no obvious sense in which the four-dimensional entity "pre-exists."   (Gibson and Pooley 2006: 191)

I concede that 'pre-existing' may be misleading in this context. There is no obvious sense in which a 4D perduring object *pre*-exists relative to its 3D parts. Perhaps 'exists' would be a better way of putting the idea. But note that it would have the same effect. The fact of the existence of 4D objects, posited by the perdurantist ontology, should not be taken lightly. After all, this is precisely what the endurantist so vehemently denies! Indeed the claim that temporally-extended perduring wholes exist must be taken as seriously as the claim that spatially extended three-dimensional objects exist. It is only if the latter exist that their 3D invariant shapes (see Figures 8.5 and 8.7) can be invoked to explain a variety of 2D perspectival projections (see Figure 8.4 and 8.6). No 3D objects—no 3D shapes, hence no 2D projections thereof. For the same reason, the unified and insightful explanation of a variety of 3D shapes (see Figure 8.8), which drives the argument from special relativity to perdurantism, is available only to someone who takes 4D objects as seriously as non-philosophers take cats, trees, and (perhaps) houses.

References

Anderson, John (1967), *Principles of Relativity Physics* (New York: Academic Press).

Armstrong, David (1980), 'Identity Through Time', in P. van Inwagen (ed.), *Time and Cause: Essays Presented to Richard Taylor* (Dordrecht: D. Reidel), 67–78.

Arntzenius, Frank (2008), 'Gunk, Topology, and Measure', in D.W. Zimmerman (ed.), *Oxford Studies in Metaphysics*, vol. 4 (Oxford: Oxford University Press), 225–47.

Arthur, Richard (2006), 'Minkowski Spacetime and the Dimensions of the Present', in Dennis Dieks (ed.), *The Ontology of Spacetime* (Amsterdam: Elsevier), 129–55.

Baker, Lynne Rudder (2000), *Persons and Bodies* (Cambridge: Cambridge University Press).

Balashov, Yuri (1999), 'Relativistic Objects', *Noûs* 33: 644–62.

—— (2000*a*), 'Enduring and Perduring Objects in Minkowski Space-Time', *Philosophical Studies* 99: 129–66.

—— (2000*b*), 'Relativity and Persistence', *Philosophy of Science* 67 (Proceedings): S549–S562.

—— (2000*c*), 'Persistence and Space-Time: Philosophical Lessons of the Pole and Barn', *Monist* 83: 321–40. Repr. in Haslanger and Kurtz 2006: 451–70.

—— (2002), 'On Stages, Worms, and Relativity', in C. Callender (ed.), *Time, Reality, and Experience* (Cambridge: Cambridge University Press), 223–52.

—— (2003*a*), 'Temporal Parts and Superluminal Motion', *Philosophical Papers* 32: 1–13.

—— (2003*b*), 'Restricted Diachronic Composition, Immanent Causality, and Objecthood: A Reply to Hudson', *Philosophical Papers* 32: 23–30.

—— (2005*a*), 'On Vagueness, 4D and Diachronic Universalism', *Australasian Journal of Philosophy* 83: 523–31.

—— (2005*b*), 'Times of Our Lives: Negotiating the Presence of Experience', *American Philosophical Quarterly* 42: 295–309. Repr. in L.N. Oaklander (ed.), *Philosophy of Time: Critical Concepts in Philosophy* (London: Routledge), 2008, vol. 3.

—— (2005*c*), 'Special Relativity, Coexistence and Temporal Parts: A Reply to Gilmore', *Philosophical Studies* 124: 1–40.

—— (2007), 'About Stage Universalism', *Philosophical Quarterly* 57: 21–39.

—— (2008), 'Persistence and Multilocation in Spacetime', in D. Dieks (ed.), *The Ontology of Spacetime*, vol. 2 (Amsterdam: Elsevier), 59–81.

Balashov, Yuri (2009), 'Pegs, Boards, and Relativistic Perdurance', *Pacific Philosophical Quarterly* 90: 167–75.

—— and Janssen, Michel (2003), 'Presentism and Relativity: A Critical Notice', *British Journal for the Philosophy of Science* 54: 327–46. Repr. in L.N. Oaklander (ed.), *Philosophy of Time: Critical Concepts in Philosophy* (London: Routledge), 2008, vol. 4.

Barbour, Julian (1999), *The End of Time: The Next Revolution in Our Understanding of the Universe* (London: Weidenfeld & Nicholson).

Belot, Gordon (2000), 'Geometry and Motion', *British Journal for the Philosophy of Science* 51: 561–95.

Bigelow, John (1996), 'Presentism and Properties', in J. Tomberlin (ed.), *Philosophical Perspectives*, vol. 10, *Metaphysics* (Cambridge, MA: Blackwell), 35–52.

Bittner, Thomas and Donnelly, Maureen (2004), 'The Mereology of Stages and Persistent Entities', in R. Lopez de Mantaras and L. Saitta (eds.), *Proceedings of the European Conference of Artificial Intelligence* (IOS Press), 283–7.

Bourne, Craig (2006), *A Future for Presentism* (Oxford: Oxford University Press).

Broad, C. D. (1923), *Scientific Thought* (New York: Harcourt, Brace and Co.).

Brown, Harvey (2005), *Physical Relativity. Space-time Structure From a Dynamical Perspective* (Oxford: Oxford University Press).

—— and Pooley, Oliver (2002), 'Relationalism Rehabilitated? I: Classical Mechanics', *British Journal for the Philosophy of Science* 53: 183–204.

Butterfield, Jeremy (1984), 'Seeing the Present', *Mind* 93: 161–76.

—— (2006a), 'The Rotating Discs Argument Defeated', *British Journal for the Philosophy of Science* 57: 1–45.

—— (2006b), 'Against Pointillisme in Mechanics', *British Journal for the Philosophy of Science* 57: 709–54.

Callender, Craig (2000a), 'Shedding Light on Time', *Philosophy of Science* 67 (Proceedings): S587–S599.

—— (2000b), 'Humean Supervenience and Rotating Homogeneous Matter', *Mind* 110: 25–43.

Carter, William and Hestevold, H. Scott (1994), 'On Passage and Persistence', *American Philosophical Quarterly* 31: 269–83.

Casati, Roberto and Varzi, Achille (1999), *Parts and Places: The Structures of Spatial Representation* (Cambridge, MA: MIT Press).

Chalmers, David, Manley, David, and Wasserman, Ryan (eds.) (2009), *Metametaphysics* (Oxford: Oxford University Press).

Craig, William Lane (2001), *Time and the Metaphysics of Relativity* (Dordrecht: Kluwer).

—— and Smith, Quentin (2008), *Einstein, Relativity and Absolute Simultaneity* (London: Routledge).

Crisp, Thomas (2005), 'Presentism and Cross-Time Relations', *American Philosophical Quarterly* 42: 5–17.

—— (2007), 'Presentism and Grounding', *Noûs* 41: 90–109.

—— and Smith, Donald (2005), ' "Wholly Present" Defined', *Philosophy and Phenomenological Research* 71: 318–44.

Donnelly, Maureen (2009), 'Parthood and Multi-Location', in D. W. Zimmerman (ed.), *Oxford Studies in Metaphysics*, vol. 5 (Oxford: Oxford University Press), 203–43.

Eagle, Antony (2009a), 'Location and Perdurance', in D. W. Zimmerman (ed.), *Oxford Studies in Metaphysics*, vol. 5 (Oxford: Oxford University Press), 53–94.

—— (2009b), 'Duration in Relativistic Spacetime', in D. W. Zimmerman (ed.), *Oxford Studies in Metaphysics*, vol. 5 (Oxford: Oxford University Press), 113–17.

—— (MS), 'Can We Read Metaphysics Off Physics? Or, What Presentists Should Say About Special Relativity'.

Earman, John (1989), *World Enough and Space-Time: Absolute Versus Relational Theories of Space and Time* (Cambridge, MA: MIT Press).

Effingham, Nick, and Robson, Jon (2007), 'A Mereological Challenge to Endurantism', *Australasian Journal of Philosophy* 85: 633–40.

Field, Hartry (1973), 'Theory Change and the Indeterminacy of Reference', *Journal of Philosophy* 70: 462–81.

Forrest, Peter (1996), 'How Innocent is Mereology?', *Analysis* 56: 127–31.

Friedman, Michael (1983), *Foundations of Space-Time Theories: Relativistic Physics and Philosophy of Science* (Princeton, NJ: Princeton University Press).

Geroch, Robert (1978), *General Relativity from A to B* (Chicago and London: University of Chicago Press).

Gibson, Ian, and Pooley, Oliver (2006), 'Relativistic Persistence', in J. Hawthorne (ed.), *Philosophical Perspectives*, vol. 20, *Metaphysics* (Oxford: Blackwell), 157–98.

Gilmore, Cody (2002), 'Balashov on Special Relativity, Coexistence, and Temporal Parts', *Philosophical Studies* 109: 241–63.

—— (2004), 'Material Objects: Metaphysical Issues'. Princeton University Dissertation.

—— (2006), 'Where in the Relativistic World Are We?', in J. Hawthorne (ed.), *Philosophical Perspectives*, vol. 20, *Metaphysics* (Oxford: Blackwell), 199–236.

—— (2007), 'Time Travel, Coinciding Objects, and Persistence', in D. W. Zimmerman (ed.), *Oxford Studies in Metaphysics*, vol. 3 (Oxford: Oxford University Press), 177–98.

—— (2008), 'Persistence and Location in Relativistic Spacetime', *Philosophy Compass* 3 (6): 1224–54.

Gilmore, Cody (2009*a*), 'Why Parthood Might Be a Four-Place Relation, and How It Behaves If It Is', in Ludger Honnefelder, Edmund Runggaldier, and Benedikt Schick (eds.), *Unity and Time in Metaphysics* (Berlin: de Gruyter), 83–133.

—— (2009*b*), 'Coinciding Objects and Duration Properties: Reply to Eagle', in D. W. Zimmerman (ed.), *Oxford Studies in Metaphysics*, vol. 5 (Oxford: Oxford University Press), 95–111.

Hales, S. D. and Johnson, T. (2003), 'Endurantism, Perdurantism, and Special Relativity', *Philosophical Quarterly* 213: 524–39.

Haslanger, Sally (1989), 'Endurance and Temporary Intrinsics', *Analysis* 49: 119–25.

—— (2003), 'Persistence Through Time', in M.J. Loux and D.W. Zimmerman (eds.), *The Oxford Handbook of Metaphysics* (Oxford: Oxford University Press), 315–54.

—— and Kurtz, Roxanne (eds.) (2006), *Persistence: Contemporary Readings* (Cambridge, MA: MIT Press).

Hawley, Katherine (2001), *How Things Persist* (Oxford: Clarendon Press).

—— (2008), 'Temporal Parts', E.N. Zalta (ed.), *The Stanford Encyclopedia of Philosophy*. URL = <http://plato.stanford.edu/entries/temporal-parts>.

Hawthorne, John (2006), 'Three-Dimensionalism', in J. Hawthorne, *Metaphysical Essays* (Oxford: Clarendon Press), 85–109.

—— and Sider, Theodore (2002), 'Locations', *Philosophical Topics* 30: 53–76.

Heller, Mark (1990), *The Ontology of Physical Objects: Four Dimensional Hunks of Matter* (Cambridge: Cambridge University Press).

—— (1993), 'Varieties of Four-Dimensionalism', *Australasian Journal of Philosophy* 71: 47–59.

—— (2000), 'Temporal Coincidence is not Overlap', *Monist* 83: 362–80.

Hinchliff, Mark (1996), 'The Puzzle of Change', in J.E. Tomberlin (ed.), *Philosophical Perspectives*, vol. 10 (Oxford: Basil Blackwell), 119–36.

Hudson, Hud (2001), *A Materialist Metaphysics of the Human Person* (Ithaca, NY: Cornell University Press).

—— (2004) (ed.), *Simples and Gunk*, *Monist* 87: No. 3.

—— (2006), *The Metaphysics of Hyperspace* (Oxford: Oxford University Press).

Huggett, Nick (1999), *Space from Zeno to Einstein* (Cambridge, MA: MIT Press).

—— (2006), 'The Regularity Account of Relational Spacetime', *Mind* 115: 41–73.

Hume, David ([1739] 1978), *Treatise of Human Nature* (Oxford: Clarendon Press).

Janssen, Michel (2002*a*), 'COI Stories: Explanation and Evidence from Copernicus to Hockney', *Perspectives on Science* 10: 457–522.

—— (2002*b*), 'Reconsidering a Scientific Revolution: The Case of Einstein versus Lorentz', *Physics in Perspective* 4: 421–46.

Johnston, Mark (1987), 'Is There a Problem about Persistence?', *Aristotelian Society* (Suppl.) 61: 107–35.

Kleinschmidt, Shieva (2007), 'Some Things About Stuff', *Philosophical Studies* 135: 407–23.

—— (ed.) (forthcoming), *Mereology and Location* (Oxford: Oxford University Press).

Koslicki, Kathrin (2003), 'The Crooked Path from Vagueness to Four-Dimensionalism', *Philosophical Studies* 114: 107–34.

Lewis, David (1983), 'Survival and Identity', in D. Lewis, *Philosophical Papers*, vol. I (Oxford: Oxford University Press), 55–77.

—— (1986), *On the Plurality of Worlds* (Oxford: Basil Blackwell).

—— (1988), 'Rearrangement of Particles: Reply to Lowe', *Analysis* 48: 65–72.

—— (1999), 'Zimmerman and the Spinning Sphere', *Australasian Journal of Philosophy* 77: 209–12.

Lombard, Lawrence (1999), 'On the Alleged Incompatibility of Presentism and Temporal Parts', *Philosophia* 27: 253–60.

Lowe, E. J. (1988), 'The Problems of Intrinsic Change: Rejoinder to Lewis', *Analysis* 48: 72–7.

MacBride, Fraser (2001), 'Four New Ways to Change Your Shape', *Australasian Journal of Philosophy* 79: 81–9.

McCall, Storrs (1994), *A Model of the Universe* (Oxford: Clarendon Press).

McDaniel, Kris (2003), 'No Paradox of Multi-Location', *Analysis* 63: 309–11.

—— (2007), 'Extended Simples', *Philosophical Studies* 133: 131–41.

McGrath, Matthew (2007a), 'Four-Dimensionalism and the Puzzles of Coincidence', in D.W. Zimmernam (ed.), *Oxford Studies in Metaphysics*, vol. 3 (Oxford: Oxford University Press), 143–76.

—— (2007b), 'Temporal Parts', *Philosophy Compass* 2 (5): 730–48.

McKinnon, Neil (2002), 'The Endurance/Perdurance Distinction', *Australasian Journal of Philosophy* 80: 288–306.

Markosian, Ned (1998), 'Brutal Composition', *Philosophical Studies* 92: 211–49.

—— (2003), 'A Defense of Presentism', in D.W. Zimmerman (ed.), *Oxford Studies in Metaphysics,* vol. 1 (Oxford: Oxford University Press), 47–82.

—— (2004a), 'Simples, Stuff, and Simple People', *Monist* 87: 405–28.

—— (2004b), 'Two Arguments from Sider's *Four-Dimensionalism*', *Philosophy and Phenomenological Research* 68: 665–73.

—— *Things and Stuff*, MS.

Maudlin, Tim (1990), 'Substances and Space-time: What Aristotle Would Have Said to Einstein', *Studies in History and Philosophy of Science* 21: 531–61.

Merricks, Trenton (1994), 'Endurance and Indiscernibility', *Journal of Philosophy* 91: 165–84.

Merricks, Trenton (1995), 'On the Incompatibility of Enduring and Perduring Entities', *Mind* 104: 523–31.

—— (1999), 'Persistence, Parts and Presentism', *Noûs* 33: 421–38.

Meyer, Ulrich (2006), 'Worlds and Times', *Notre Dame Journal of Formal Logic* 47: 25–37.

Miller, Kristie (2004), 'Enduring Special Relativity', *Southern Journal of Philosophy* 42: 349–70.

Minkowski, Hermann ([1908] 1952), 'Space and Time', in H.A. Lorentz, A. Einstein, H. Minkowski, and H. Weyl, *The Principle of Relativity* (New York: Dover), 73–91.

Monton, Bradley (MS), 'Prolegomena to Any Future Physics-Based Metaphysics'. URL = <http://philsci-archive.pitt.edu/archive/00004094>

Moyer, Mark (2008), 'Why We Shouldn't Swallow Worm Slices: A Case Study in Semantic Accommodation', *Noûs* 42: 109–38.

Nerlich, Graham (1994), *What Spacetime Explains* (Cambridge: Cambridge University Press).

Nolan, Daniel (2006), 'Vagueness, Multiplicity and Parts', *Noûs* 40: 716–37.

Norton, John (2008a), 'Why Constructive Relativity Fails', *British Journal for the Philosophy of Science* 59: 821–34.

—— (2008b), 'The Hole Argument', in Edward N. Zalta (ed.), *The Stanford Encyclopedia of Philosophy* (Winter 2008 ed.), URL = <http://plato.stanford.edu/archives/win2008/entries/spacetime-holearg/.

Oderberg, David (1993), *The Metaphysics of Identity over Time* (New York: St. Martin's Press).

Parsons, Josh (2000), 'Must a Four-Dimensionalist Believe in Temporal Parts?', *Monist* 83: 399–418.

—— (2007), 'Theories of Location', in D.W. Zimmerman (ed.), *Oxford Studies in Metaphysics*, vol. 3 (Oxford: Oxford University Press), 201–32.

Petkov, Vesselin (2005), *Relativity and the Nature of Spacetime* (Berlin: Springer).

Prior, A.N. (1957), *Time and Modality* (Oxford: Clarendon Press).

Pryce, M.H.L. (1948), 'The Mass-Centre in the Restricted Theory of Relativity and Its Connexion with the Quantum Theory of Elementary Particles', *Proceedings of the Royal Society of London* 195A: 62–81.

Putnam, Hilary (1975), 'Philosophy and our Mental Life', in H. Putnam, *Mind, Language, and Reality* (Cambridge: Cambridge University Press), 291–303.

Quine, W.V. (1953), 'Mr Strawson on Logical Theory', *Mind* 62: 433–51.

—— (1960), *Word and Object* (Cambridge, MA: MIT Press).

—— (1963), 'Identity, Ostension, and Hypostasis', in W.V. Quine, *From a Logical Point of View* (Evanston: Harper & Row), 65–79.

—— (1987), 'Space-Time', in W.V. Quine, *Quiddities* (Cambridge, MA: Harvard University Press), 196–9.

Rea, Michael (ed.) (1997), *Material Constitution* (Lanham, MD: Rowman & Littlefield Publishers).

—— (1998), 'Temporal Parts Unmotivated', *Philosophical Review* 107: 225–60.

Rindler, Wolfgang (1961), 'Length Contraction Paradox', *American Journal of Physics* 29: 365–6.

Russell, Bertrand (1927), *The Analysis of Matter* (New York: Harcourt, Brace & Co.).

Russell, Jeffrey T. (2008), 'The Structure of Gunk: Adventures in the Ontology of Space', in D. W. Zimmerman (ed.), *Oxford Studies in Metaphysics*, vol. 4 (Oxford: Oxford University Press), 248–74.

Sartori, Leo (1996), *Understanding Relativity* (Berkeley, CA: University of California Press).

Sattig, Thomas (2006), *The Language and Reality of Time* (Oxford: Clarendon Press).

Saucedo, Raul (forthcoming), 'Parthood and Location', in D.W. Zimmerman (ed.), *Oxford Studies in Metaphysics*, vol. 6 (Oxford: Oxford University Press).

Saunders, Simon (2002), 'How Relativity Contradicts Presentism', in C. Callender (ed.), *Time, Reality, and Experience* (Cambridge: Cambridge University Press), 277–92.

Savitt, Steven (2000), 'There's No Time Like the Present (in Minkowski Space-time)', *Philosophy of Science* 67 (Suppl.): S563–S574.

—— (2009), 'The Transient *Nows*', in Joy Christian and Wayne C. Myrvold (eds.), *Quantum Reality, Relativistic Causality, and Closing the Epistemic Circle: Essays in Honour of Abner Shimony* (Berlin: Springer), 349–68.

Schaffer, Jonathan (2009), 'Spacetime the One Substance', *Philosophical Studies* 145: 131–48.

Schattner, R. (1978), 'The Center of Mass in General Relativity', *General Relativity and Gravitation* 10: 377–93.

Sider, Theodore (2000), 'The Stage View and Temporary Intrinsics', *Analysis* 60: 84–8.

—— (2001), *Four-Dimensionalism: An Ontology of Persistence and Time* (Oxford: Clarendon Press).

—— (2003), 'Against Vague Existence', *Philosophical Studies* 114: 135–46.

—— (2008), 'Temporal Parts', in T. Sider, J. Hawthorne, and D. W. Zimmerman (eds.), *Contemporary Debates in Metaphysics* (Oxford: Blackwell), 241–62.

Simons, Peter (1987), *Parts: A Study in Ontology* (Oxford: Oxford University Press).

—— (2004), 'Extended Simples: A Third Way Between Atoms and Gunk', *Monist* 87: 371–84.

Sklar, Lawrence (1974), *Space, Time, and Spacetime* (Berkeley, CA: University of California Press).

Skow, Brad (2007), 'Are Shapes Intrinsic?', *Philosophical Studies* 133: 111–30.

Smart, J.J.C. (1963), *Philosophy and Scientific Realism* (London: Routledge & Kegan Paul).

—— (1972), 'Space-Time and Individuals', in R. Rudner and I. Scheffler (eds.), *Logic and Art: Essays in Honor of Nelson Goodman* (New York: Macmillan), 3–20; repr. in Smart 1987: 61–77.

—— (1987), *Essays Metaphysical and Moral: Selected Philosophical Papers* (Oxford: Blackwell).

—— (2008), 'The Tenseless Theory of Time', in T. Sider, J. Hawthorne, and D. W. Zimmerman (eds.), *Contemporary Debates in Metaphysics* (Oxford: Blackwell), 226–38.

Smith, Quentin (1993), *Language and Time* (New York: Oxford University Press).

Sorkin, Rafael (1991), 'Spacetime and Causal Sets', in J.C.D'Olivo et al. (eds.), *Relativity and Gravitation: Classical and Quantum*, Proceedings of the SILARG VII Conference, Cocoyoc, Mexico, Dec. 1990 (Singapore: World Scientific), 150–73.

Stein, Howard (1991), 'On Relativity Theory and Openness of the Future', *Philosophy of Science* 58: 147–67.

Swoyer, Chris (1984), 'Causation and Identity', in P. French, T. Uehling, Jr., and H.K. Wettstein (eds.), *Midwest Studies in Philosophy*, vol. 9 (Minneapolis, MN: University of Minnesota Press), 593–622.

Taylor, Richard (1955), 'Spatial and Temporal Analogies and the Concept of Identity', *Journal of Philosophy* 52: 599–612.

Thomson, J.J. (1983), 'Parthood and Identity Across Time', *Journal of Philosophy* 80: 201–20.

Tooley, Michael (1997), *Time, Tense, and Causation* (Oxford: Oxford University Press).

Torretti, Roberto (1983), *Relativity and Geometry* (Oxford and New York: Pergamon Press).

van Inwagen, Peter (1990*a*), 'Four-Dimensional Objects', *Noûs* 24: 245–55.

—— (1990*b*), *Material Beings* (Ithaca, NY: Cornell University Press).

Whitehead, A.N. (1920), *The Concept of Nature* (Cambridge: Cambridge University Press).

Winnie, John (1977), 'The Causal Theory of Spacetime', in John Earman, Clark Glymour, and John Stachel (eds.), *Foundations of Space-Time Theories* (Minneapolis, MN: Minnesota University Press), 134–205.

Zimmerman, Dean (1996*a*), 'Persistence and Presentism', *Philosophical Papers* 25: 115–26.

Zimmerman, Dean (1996*b*), 'Could Extended Objects Be Made Out of Simple Parts? An Argument for "Atomless Gunk" ', *Philosophy and Phenomenological Research* 56: 1–29.

—— (1997), 'Immanent Causation', in J. Tomberlin (ed.), *Philosophical Perspectives*, vol. 11 (Oxford: Blackwell), 433–71.

—— (1998*a*), 'Temporary Intrinsics and Presentism', in D. Zimmerman and P. van Inwagen (eds.), *Metaphysics: The Big Questions* (Cambridge, MA: Blackwell), 206–19.

—— (1998*b*), 'Temporal Parts and Supervenient Causation: The Incompatibility of Two Humean Doctrines', *Australasian Journal of Philosophy* 76: 265–88.

—— (1999), 'One Really Big Liquid Sphere: Reply to Lewis', *Australasian Journal of Philosophy* 77: 213–15.

—— (2008), 'The Privileged Present: Defending an "A-theory" of Time', in T. Sider, J. Hawthorne, and D. W. Zimmerman (eds.), *Contemporary Debates in Metaphysics* (Oxford: Blackwell), 211–25.

—— (forthcoming), 'Presentism and the Space-Time Manifold', in Craig Callender (ed.), *Oxford Handbook of Time* (Oxford: Oxford University Press).

Index

absolute chronological precedence 24, 68, 71–2, 90

absolute simultaneity: *see* simultaneity: absolute

absolute time: *see* time: absolute

absurdity thesis 168, 173–5, 181

achronal part: *see* part: achronal

achronal parthood: *see* parthood: achronal

achronal region: *see* region: achronal

achronal region: *see* region: achronal

achronal slice (of an object's path in spacetime): *see* slice (of an object's path in spacetime): achronal

achronal universalism: *see* universalism: achronal

adverbialism 19, 74n, 75, 93n

affine: structure 52; connection 97n

Alexandrov-Stein present 143–8, 150–3, 155

Anderson, John xvn

Armstrong, David 14n, 31n, 78n

Arntzenius, Frank 7n, 21n

Arthur, Richard 144n, 145n, 156n

AS-coexistence: *see* coexistence: Alexandrov-Stein

AS-co-presence: *see* co-presence: Alexandrov-Stein

AS-present: *see* Alexandrov-Stein present

asymmetry thesis 168–73, 179

A-theory(ies) of time: *see* time: A-theory(ies) of

atomism 6–7

Baker, Lynne Rudder 14n

Barbour, Julian 5n

Belot, Gordon 5n

Bigelow, John 2n

Bittner, Thomas 14n, 16n, 23n, 26n, 36n

Bourne, Craig 2n

Broad, C. D. 2n

Brown, Harvey 4n, 5n, 58n

B-theory(ies) of time: *see* time: B-theory(ies) of

Butterfield, Jeremy 6n, 7n, 21n, 31n, 78n, 145n

Callender, Craig 2n, 31n, 78n

Carter, William 103n

Casati, Roberto 11n

CASH (Coexistence As Sharing a Hyperplane of simultaneity) 138–43, 145, 147–8, 151–5, 158, 159–60, 164, 167–8, 171–6, 179, 181–2, 188–9

CASS (Coexistence As Spacelike Separation) 139n

Cauchy surface 25

Chalmers, David 10n

coexistence:

   Alexandrov-Stein 143–55, 157, 161–2

   as a multigrade relation 131, 134, 138, 140–1, 148, 174, 176–89

   chronological incoherence of 186–9

   contextuality of 181–6, 189

   in classical (Galilean) spacetime 132, 135–8, 158–9, 176–9

   in Minkowski spacetime 108–109, 132, 138–58, 179–90

   in spacetime 131–96

   relative to a hyperplane of simultaneity 134n

   temporally laden determinations of 137, 164, 168, 170–3, 176–7

coexistence*:

   Alexandrov-Stein 146–8, 161–2

   in classical (Galilean) spacetime 136–8, 158–9

   in Minkowski spacetime 139–42, 146–7, 159–62, 168–72, 174–6, 179, 181, 182n, 183

coexistence**:

   Alexandrov-Stein 146–8, 161–2

   in classical (Galilean) spacetime 137–8, 158–9